Food Famine and Fertilizers

FOOD FAMINE AND FERTILIZERS

By

Seshadri Kannan

A.P.H. PUBLISHING CORPORATION
4435-36/7, ANSARI ROAD, DARYA GANJ
NEW DELHI-110 002

Published by
S.B. Nangia
A P H Publishing Corporation
4435-36/7, Ansari Road, Daryaganj
New Delhi 110002
Ph.: 23274050
E-mail : aphbooks@gmail.com

2024

Rs 3995/-

Printed at
Balaji Offset
Navin Shahdara, Delhi 110032

PREFACE

"What might head a list of the defining characteristics of the human species? While our view of ourselves could hardly avoid highlighting our accomplishments in engineering, art, medicine, space travel and the like, in a more dispassionate assessment agriculture would probably displace all other contenders for top billing".

"Most of the other achievements of humankind have followed from this one. Almost without exception, all people on earth today are sustained by agriculture. With a minute number of exceptions, no other species is a farmer. Essentially all of the arable land in the world is under cultivation. Yet agriculture began just a few thousand years ago, long after the appearance of anatomically modern humans". (Wadley G, Martin A. 2000) (Journal of Australasian College of Nutritional and Environmental Medicine 19, 3–12)

Food, clothing and shelter are the basic requirement of mankind. The first one food, keeps the organism to survive, grow and multiply or expand. Humans due to the constant evolution have the capacity to search, select and produce food necessary for growth and survival. From a nomadic life with the hunting and feeding on the kill, humans have developed the skill to change the foraging to a farming habitat. The book is a collection of information obtained from many scientific articles and magazines as also some known and a less known historical and scientific facts relating to the three areas viz., food, famine and fertilizers. These topics are related to each other and are very important today when food is becoming a limiting factor, and famine, latent or imminent, threatens human survival through malnutrition, diseases and eventually political unrest. Fertilizers have to be recognized as 'necessary evil', because, but for it, the 'green revolution' would never have been successful, and at the same time, excessive fertilization causes ground and river water pollution, and has become the evil by replacing organic fertilizers, which were in use for centuries. The information presented here relates to the problems in many countries, but not exhaustive. Food shortages and famines were occurring over several periods of time, and the struggle for survival also is a continuous process. Industrialization has greatly helped in the development of food production, and at the same time has been the cause of degradation of soil, water and air. Some of these areas have been discussed as related to the three topics. Green Revolution temporarily put off the acute starvation. History reveals there have been revolutions in different forms. 'Revolution' is only a universal exhibition of human efforts for survival and as the population increases there is a greater need for discoveries in plant science. Many of these are discussed. The benefits of evolving GM crop species are also presented in different chapters.

The permissions granted by the editor of Tehelka Dr. Satish Warrier, http://www.tehelka.com, "A Sham in the Name of Indira Gandhi June 18 , 2005"; by Stanford News Service (Mark Shwartz) on soil erosion in Kenya);"David Stuart-CAPITA" <david.stuart@bbc.co.uk> text from the (http://www.bbc.co.uk/history/british/empire_seapower/agricultural_revolution_01.shtml) BBC history website; Dr. Dennis Crowley, CEO <dennis@acnem.org> and Dr. Michelle Bradford, ACNEM, Black Rock, Victoria 3193, Australia for quotes from Journal of Australasian College of Nutritional and Environmental Medicine 19, 3–12; Dr. Ellie Smith (Permissions@theatlantic.com) for the lines from the article by Charles C. Mann <http://www.theatlantic.com/doc/by/charles_c_mann>; are gratefully acknowledged.

The information in the articles referred to in *California Agriculture* are acknowledged and credited to: *California Agriculture*, 38: Oct ©1984 p 2; 40, May June, pages 2-3, ©1986, 40:2-3, University of California; The Regents of the University of California. These belong to public domain and are free for quoting, with due acknowledgement.

Photo (pages 6-7, for wireless watering) by James Kim and provided by U.S. Dept. of Agriculture, Agricultural Research Service, Sidney, Montana, USA. in Agricultural Research, 55, No.6, July 2007 pp.6-7 and the picture in page 10, in the same issue are gratefully acknowledged. The author has included excerpts from the articles by R.M. Welch, deQuattro and D. Comis in the May and December issues of 1999, and R. Evans (July:6-7), J. Teasdale (July: 4-5) , L. McGinnis (July:8-11), in vol 55,2007, Agricultural Research journals and wishes to acknowledge with thanks.

Finally, although I did the job of preparing the book, the final touch in the form of giving a good shape and arrangement has been beautifully done by the Publisher, Dr. Nangia and I thank him for this and for getting the book published in an elegant manner.

Seshadri Kannan

38 South Mada St
Thiruvanmiyur
Chennai 600041

CONTENTS

CHAPTER-I

FOOD

1. HISTORICAL

Perhaps the most significant single event in human history is the development of agriculture and animal husbandry. Humans evolved themselves from food-gathering nomads to food-producing social groups which took place during several millennia between 10,000 to 16,000 years ago and contemporarily in several regions and several places where people developed colonies and lived together. This period is recorded as "Neolithic Revolution", a term which was first introduced by Childe in 1928 (see Wirtz and Lemmen, 2003), in which abundant food, and sedentary way of life was the forerunner of the expanding population. While the Paleolithic period extends from 2 million years to ten thousand years, the very short Neolithic period from 5,500 to 2,500 BC covers the time with the onset of farming and ending the widespread use of metal tools. However, there is no clear demarcation of the period within this Neolithic time since it differed with regions. It appears that this term was introduced in the 1930s, but some writers prefer to call it 'agricultural revolution' and thus it may be the first of such revolution which has taken place several thousand years ago. The real advancement in farming began a few centuries before the Industrial Revolution, although no precise dating is available, as also any archeological evidence.

Writing about the ancient history in the northeast of Mexico, Múzquiz (2000) states the Canyon del Diablo in the state of Tamaulipas and in Mexico is one of the most important sites from the point of anthropological and archeological studies unraveling the history of humans changing from the hunting and gathering habitat to farming, viz., learning to produce food, and leading a sedentary life. The inhabitants of Tamaulipas got the benefit of cultivating plants like corn, which were domesticated around 2500 BC by the earlier groups. The Tamaulipas inhabitants left behind some important archeological clues- the rupestrian (cave) paintings in the Cañón del Diablo. The American archeologist, Richard S. MacNeish carried out important archeo-paleo-botanical work on the rock shelters and other archeological sites during 1946 and 1954 while trying to shed light on the development of agriculture and the origins of corn on the American continent. On the basis of his findings he classified the cultural phases into nine groups, the earliest one as the most primitive *Dablo* phase, and the last one ending with the Los Angeles phase in 1748.

Flannery and Marcus (2001) wrote in the National Academy Press about MacNeish and his interest in finding out answers through his excavation, for his two questions: how maize was first domesticated and how it influenced Mesoamerican civilization. He chose the village of Puebla in Mexico on the reason that the climate being dry there helped preserve the past. His excavation in Tamaulipas area which covered about 450 sites spread over 12 caves, revealed that corn had been domesticated long before. Through radiocarbon dating, the period was fixed to over 12,000 years, and that the people belonged to a presedentary era; they traveled in small bands as they were gathering their food. The plant and animal remains of the area were suggestive of their availability at different seasons in a year. From the subsequent examination of different caves, he found evidence which pointed to habitation of the region around 40,000 years ago which nearly would contradict the long standing theory that human life in the New World was only about 11,000 years old. The evidence from Pendejo Cave will force a re-thinking of early man in the America's, "and will place his time between 40,000 and 50,000 years" according to MacNeish (Flannery and Marcus, 2001). MacNeish had occasion to examine some caves in China and in 1993 he and his team began the excavation of Xian Ren Dong ("Benevolent Spirit Cave") and Wang Dong ("Bucket Handle Cave"). In their preliminary survey they found 'phytoliths' of wild rice *Oryza nivara* and the period was fixed as 17,040 BP (Before the Present). The first phytoliths of domestic rice *Oryza sativa* obtained in both the caves were identified as belonging to 14,000 and 11,200 BP. although this rice became dominant during 9600 to 8000 B.P. the period marking the advent of cereal agriculture in the Near East.

2. ORIGIN OF AGRICULTURE

Domestication and Neolithic Revolution

The origin of agriculture has been the subject of search and analysis by archeologists, paleontologists and plant biologists all over the world. One of the earliest reviews on the origin of agriculture was that of Flannery (1973). With extensive discussion on the theories on the origin of agriculture in four regions, namely Southwest Asia, Southeast Asia, Mesoamerica, and the Andes, he concluded that there is no one model that could explain the origin in all these regions together. Furthermore, cultivation might have begun after the establishment of village life in some parts of the Near East and Peru and it must have also happened in the Mesoamerica region while the people were still nomadic with their hunting and gathering food habits continuing for thousands of years even after they began farming. Flannery further discusses the scope whether there were exchanges of crops amongst the four centers, in the early periods. The most important ethnobotanical question he raises is: "How independent was early agriculture in Asia, Africa, and the Americas? Could one area have influenced the others by diffusion of early crops?" He relates to an earlier article by Pickersgill and Bunting (1969) who had examined the botanical evidence for four plants namely, cotton, bottle gourd, coconut and sweet potato in connection with the diffusion theory. According to them, the bottle gourd would not have been introduced to the New World from Africa or Asia and would have reached South America long before man could introduce it there. The sweet potato (*Ipomea batatas*) might have arisen by hybridization between a diploid and a tetraploid ancestor since both were present in Polynesia and the Americas and crossing might have occurred there to produce the hexaploid sweet potato.

The early domestication of plants and animals and the transition of the early inhabitants from foraging to farming are said to have taken place independently in different regions of the world. It is important to note that the discovery of 'fire' was the first weapon of destruction if one may call so, and but for it, humans would not have been able to keep the wild animals at bay, and perhaps would have continued to be 'cave' men. Further, the humans having evolved from monkeys over several millennia before, did not have the 'claws' and 'paws' which would help 'kill' animals for food; and therefore, must have been 'vegetarians' living on available plant foods, perhaps learning from the early ancestors. Sir Arthur C. Clarke, in his fiction "2001 A Space Odyssey", wrote that the early cave-dwelling humans were 'moon-gazers' and accidentally rolled a big stone on a pig which got killed, and then the humans learnt to eat meat. Perhaps there may be many occasions such efforts would have happened before humans turned to hunting and foraging.

According to Smith (1997) who has reviewed the work of several others, three major crop plants, namely, maize (*Zea mays*), the common bean (*Phaseolus vulgaris*) and squash (*Cucurbita pepo*) were first domesticated in Mesoamerica and various techniques including the radiocarbon age determination including small sample accelerator mass spectrometer radiocarbon method reveal that the domestication occurred more recently than 7000 to 10,000 years ago as was thought earlier. Smith studied various characteristics of the squash seeds, peduncles and fruit rind fragments obtained in the Guilá Naquitz cave in Oaxaca, Mexico, and concluded that the cultivation of *Cucurbita pepo* by the inhabitants of the cave, was between 10,000 to 8000 years ago, which precedes by 4000 years the domestication of maize and beans.

There are more recent reviews on the shift of humans from hunting to food gathering or primitive agriculture, known as 'Neolithic agriculture.' (Weisdorf, 2005). Some of the views of Weisdorf (2005) are summaarised here. First they had domesticated many plants and animals perhaps unknowingly as a prelude to farming. But the question remained as to what was the driving force for these humans to resort to farming. Archaeologists, agronomists, anthropologists and historians made several hypotheses relating to this transformation. One such hypothesis suggests that with the extinction of animals due to the large scale hunting by Paleolithic humans there arose the need for farming and development of agriculture.

With the increase in population there might have been a change in their lifestyles from foraging to sedentary colonization. Weisdorf (2005) mentions the ancient Greeks recorded three stages of development , viz., first came hunting and gathering, then domestication of animals and nomadic life style, and the final one the invention of 'agriculture'. This is called the 'stage' hypothesis as described by Isaac (*see* Weisdorf). The 'oasis' hypothesis, 'marginal zone' hypothesis and 'population' hypothesis were proposed by a few others at different times. The oasis hypothesis explains that the shift was motivated by changes in the environment, but this was not accepted for the reason that climate changes were too slow and no shift took place when there were environmental change during the interglacial periods. The 'overkill' hypothesis explains the shift resulting from animal extinction and consequent 'food crisis'. This is also not accepted because animal extinction and farming did not occur together or in a sequence. The reason that population pressure provided the impetus for the shift to farming was accepted by historians and archeologists during 1970s. Weisdorf quotes Cohen's hypothesis that population growth 'spikes' occurred frequently throughout human history and that led to over-population as early as 15,000 years ago. The stress brought about by the increasing population and the depletion of food resources had obviously made the humans move to areas of foods of greater abundance. This reasoning was in accord with the fact that the earlier humans started migrating from Africa which was the original home for *Homo sapiens*, and moved to more inhabitable areas in Europe, Asia and even the New World.

Mithen (1996) considers that the early humans possessed knowledge about the propagation of animals and plants but did not know how to domesticate them. He further explains that the origins of agriculture which took place several thousand years ago were due to a fundamental change in the conception of the human mind with regard to farming in particular. In other words, if humans had not developed this capacity, the change from foraging to farming would not have taken place. Although various hypotheses were discussed, Weisdorf concludes that there has not been any single model, which has received widespread acceptance amongst the investigators. As new evidence on this transition is constantly appearing in literature, the theories also get modified. He also records that sedentism occurred earlier to the transition to agriculture and the tools for farming were also available then, which helped the foragers to take to agriculture. Furthermore, the egalitarian nature of the societies was replaced by hierarchical one which eventually led to an organization concentrating on farming and the household patterns.

Wirtz and Lemmen (2003) have employed mathematical modeling to study the Neolithic transition of the early humans and examined several maps of potential vegetation at 5000 BP. They have introduced terms like 'Subsistence Intensity'(SI) which is described by them as the effectiveness of a community in generating consumable foods and secondary products, and it is achieved under an agricultural-pastoral or a hunting-gathering life-style of the humans. This development is said to have occurred in different gradations in the various world regions. The 'Neolithic Revolution' perhaps triggered by environmental changes have occurred independently in at least five places. According to Smith (1998, *see* Wirtz and Lemmen, 2003), there are about five independent farming centers where global agriculture originated, and these are *Fertile Crescent*, some areas of North China and South China, Central Mexico and the central Andes. Further, the western Mediterranean region had adapted to agricultural life very early. Excluded from the list are areas where local species were domesticated first built the concept of domestication and/or other founder crops had presumably been introduced from elsewhere, e.g., The African Sahel, the Indus Valley and eastern North America.

Wirtz and Lemmen in their model described economic adaptation, growth and human population as also the spread of their cultural characteristics which accounted for over-exploitation of natural resources, high mortality rates and the climate changes over several thousand years. They have discussed various problems namely, the lag between domestication and fully developed farming, the leveling off, of the technological thrust resulting from the transition, the emergence of migration patterns of the humans and sensitivity to climate changes.

The transition from HG (hunting-gathering) to AP (agricultural pastoral) societies developed at different speeds across the continents. For the *Fertile Crescent*, the abrupt shift occurred at 9 kyr BP. The domestication of *Triticum sp.* was successfully accomplished around 12.5 kyr BP, while einkorn wheat was domesticated in northern Syria around 10.5 kyr BP. Cereal cultivation might have begun even earlier perhaps in the late Pleistocene period. The domestication of animals like goat and sheep followed after a lag period. The point of this transition was noticeable at *Catal Hoyuk*, the oldest Neolithic town known, where both HG and AP means of livelihood co-existed around 10 kyr BP (Balter, 1999). In contrast, the domestication in the New World centers viz., Ecuador and Central Mexico, took place after a gap of few millennia which is attributed to a 'domestication potential' and then the 'domestication speed'. The timing difference between the Neolithic shift in the New and Old World provides evidence for the diversification of the domestication process on a global scale, and the time lag is estimated as 3.5 kyr. (Wirtz and Lemmen, 2003). In their final conclusion they confirmed the view of Boserup (1981) that mere population pressure alone was not responsible, but a steady innovation and competition between subsistence strategies served as the prime movers for the agricultural development.

It is recorded that over 80 percent of mankind's food is provided by the grains of a dozen plant species and over the succeeding generations man invented new machines and developed techniques to increase the number of crop varieties and their yields. The following will be an overview of the history of farming. Innovation of plough and then the tractor were responsible in the farming efficiency. Early farming began in the areas of present day Turkey and the Middle East about 10,000 years ago perhaps in two of the settlements known as *Catal Hüyük* and *Jericho*. *Catal Hüyük* settlement had over thousand houses by 6000 B.C. Archeologists have found evidence of people taking wild grasses and using the seeds for food and also saving some for raising next years' crop (Rymer, 2006). They also identified the seeds as cereals which make up a large percentage of the world's food supply today. Rymer (2006) has further described about early rise of civilization. *Jericho*, like many early cities, was located around a water source, a spring that produced over 1000 gallons of water every minute. The settlement consisted of about eight to ten acres land and supported about two to three thousand people who grew wheat, barley, peas, and lentils. The early settlers in this area were hunter-gatherers who would live off the land growing berry and edible plants, as well as hunting wild animals. They lived together in smaller groups and moved to different places in search of more food. Permanent settlements came into existence as at *Jericho* when men began to plant and harvest crops in the land where they lived.

An analysis of the factors which influenced the early people to change from the hunting and food gathering, to farming leads to interesting facts. While it should have been easier to hunt the animals for food, growing crops is a long way off the normal development. The driving force is likely to be a decrease in the animals due to over-killing. Next easier way is to eat the readily available plant foods like fruits and even nuts like groundnut, which however needs knowledge of its underground occurrence and then digging the ground for lifting it up. On the other hand, even collecting the grains of cereals or any other grass, dehusking if needed for consuming the inside grains as such which are soft, and then making food from them is more labor-involving than harvesting the fruits from trees or shrubs. For all these, the development of brain, and capacity for thinking, and organizing a society are all pre-requisites for cultivating an agricultural crop. These changes may not have taken place in a short period but, several centuries of human evolution, and also simultaneously in several regions. At the same time, a sedentary life-style can only lead to such evolution in food habits. It is interesting that the people were well organized in the Harappan culture about 3000 years BC when small towns were built to accommodate a few thousand people.

3. CROPS IN THE OLD AND NEW WORLDS

Before it became the New World, the Western Hemisphere was vastly more

populous and sophisticated than has been though an altogether more salubrious place to live at the time than, say, Europe. New evidence of both the extent of the population and its agricultural advancement leads to a remarkable conjecture: the Amazon rain forest may be largely a human artifact. (Charles C. Mann) <http://www.theatlantic.com/doc/by/charles c mann>

With regard to the domestication of plants and the development of agriculture as a whole, one has to look into the agricultural evolution in both the Old and New World. Many historians believed for a long time that the Old World remained in isolation from the New World. It was realized later that it was incorrect. According to MacNeish (1964) the development of civilization began in the Mexican valley of Tehuacán, where bands of hunters became urban craftsmen in the course of 12,000 years and there might have been 'cultural migration' during this period. The European adventurers reaching the New World in the 16th century were greatly surprised to find cultures almost as advanced as theirs, in many fields. In the regions extending from Mexico down to Peru, large variety of domesticated plants, which were new, were discovered, as also agriculture more advanced than found in the Old World. The 'farmers' of these regions cultivated white and the sweet potato, vanilla and pepper (*Capsicum*) and vegetables and fruits like tomato, avocado, pineapple, guava and papaya. There was also evidence that the natives smoked tobacco in pipes and used some form of cane-cigarettes, stuffed at one end with fibers to serve as a filter, and that the credit to the present day filter cigarettes should be given to them (MacNeish, 1964). There are many descriptions about the New World civilizations. Native Americans had domesticated many plants and raised them for human food. They were originally big game hunters and while hunting, followed the animals across an "ice bridge" connecting the Bering Strait. Later they segregated into different cultural groups and developed their own unique agricultural styles. The South American Natives of the Olmec, Aztec, Inca, and Maya all had knowledge of agriculture and moved in search of their food. Plants were cultivated in many of the 'present day' regions, Mexico, Guatemala and Honduras, and one of the main crops was maize although squash, pumpkin, gourd, chili peppers, and avocados were also grown.

Historically, the early civilizations in the world were along major river systems. Egyptians settled along the Nile River, Harappa culture along the Indus, Chinese Empire along the Huang River and the Mesopotamian Countries along the rivers Tigris and Euphrates (Rymer). The main reason was that the river systems provided them plenty of water and land with soil enriched by silt brought by the yearly floods. Although they were not possibly aware, the silt from the river provided natural fertilizer, and minerals, both organic and inorganic. Farming, based predominantly on wheat and barley, first appeared in the Middle East and spread quickly to western Asia, Egypt and Europe. The early civilizations all relied primarily on cereal agriculture. Cultivation of fruit trees began three thousand years later, again in the Middle East, and vegetables and other crops followed.

Cultivation of rice began in Asia about 7000 years ago. Harappa culture was very organized, though its original place is not known, the period is said to be around 3000 BC. The people of this culture settled on the Indus Plain, the land consisting layers of alluvial deposits very rich in fertility. Addition of silt came down the Indus river in the annual floods, and was estimated to be twice the amount of what came down the Nile in Egypt. Harappans built large grain storage areas, dams and drainage systems for the farming land. They grew crops like wheat, barley, sesame and dates (Wadley and Martin, 2000).

Araus *et al.* (2007) have described the development of dryland agriculture and wheat is considered to be the principal crop brought under cultivation by the western agriculture. This seems to have taken place around 10,000 BC in the Mediterranean savannahs and steppes of the Near East, which have had long period of drought considered ideal for growing annuals having large seeds. Because of these properties the plants could survive long dry periods and would germinate no sooner than the onset of rains. Araus group also analysed the carbon and nitrogen stable isotope compositions and the dimensions of the fossil grains of wheat (*Triticum aestivum/durum*) to determine the geological time of its cultivation. The high ^{15}N values of the fossil kernel samples provided evidence that soils in early agriculture were also fertile, with high organic matter content and nitrogen derived from it. High soil fertility was maintained by leaving the land fallow between two crops. The authors inferred from their fossil analysis that the high yields obtained in the early periods were similar to the average yields obtained at the beginning of the 20th century. Furthermore they also believed that the ratio of people to land during the Neolithic period and at the present time is nearly constant, i.e., at Neolithic age some 4 to 10 million, to one billion people now were due to an enlargement of the total cultivated land, and not an increase in yield of grain per unit of land.

Recently a new method has been introduced for tracing the origin of certain crops in the New World. The Central American isthmus was considered a major route for dispersal of plant species originally under cultivation in the domestication centers of South Mexico and northern South America. Dickau *et al.* (2007) examined the starch grains recovered from the stone tools used by the inhabitants of present day Panama to process the grains, and obtained new evidence for the spread of New World 'cultigens', like maize (*Zea mays*), manioc (*Manihot esculenta*), and arrowroot (*Maranta arundinacea*). Maize starch recovered from the lowlands of central Panama confirmed that maize was cultivated for food by 7800-5600 cal BP. Examination of the starch from the humid premontane forests in the western Panama also revealed that maize and root crops were used by the people by 7400 to 5600 cal BP, the period going back to several millennia earlier than previously recorded. Dickau *et al.* (2007) obtained evidence for the use of several local carbohydrate-rich plant products from plants like *Zania* and *Dioscorea* spp. and proposed that crop dispersals took place *via* diffusion or exchange of germplasm by the people, and not through the migration of those who were raising these crops. They also discovered that three domesticated crop species, namely, maize, manioc and arrowroot, were introduced in the Panama regions which had different climatic regimes than in their original homes.

4. NATIVE INDIAN AND THE CORN PLANT

Most if not all the cereal crops are efficient trappers of solar energy and this energy is stored in the form of carbohydrates in the grains. Maize is one such crop, but perhaps the most efficient trapper of the solar energy and the atmospheric carbon dioxide; the grains of this plant provide energy-rich food for many people. In the famous book, "Hunger Fighters", Paul de Kruif (1928) begins his chapter on "The Maize Finders" thus, "What fighters of hunger can compare to the unknown finders of the American maize? Unlike wheat, unlike any food giving life to men, there is no wild maize: ...It can't grow free.... A thousand times more domestic is the maize than the most helpless goldfish: without endless storing, hoeing, reaping, saving, and sowing by men, this plant would die out—utterly." Kruif describes the early breeding work of George Harrison Shull and Henry Wallace, of the late 1800s, and Shull is considered "as the first white *breeder* of maize".

The transformation of the "wild grass", the American Indian corn *Zea mays*, was accomplished by the peoples of the New World similar to the development of wheat, barley, rye, oat and millet, from the native grasses, in the Old World. Furthermore the pre-Columbian people from Chile to the valley of St. Lawrence in Canada, used one or more of the 150 varieties of corn for their food, and corn remains even today as one of the most important grain or fodder crop in the Americas. Early historians believed this might provide the clue for the origin of the New World civilization. There is a lot of evidence that there have been human intervention in the domestication and development of crop plants since they felt the need for increasing food production in view of the rising population. MacNeish (1964) carried out excavations in the caves, in the valley of Tehuacán and discovered wild remains of corn. The corncobs were only about 20 millimeters long, and an examination revealed the presence of miniature ears of corn. These cobs about 7,000 years old, is believed to be the ancestor of today's corn. The fact that the cultivated corn would not have survived without man's intervention has been known for a long time. MacNeish believes that the people of Tehuacán wandered the valley in search of animal foods like rabbits, and small animals during the period from 12,000 and 7000 B.C. and from around 6700 B.C. until about 5000 B.C. the pattern changed into a new phase, the hunters becoming collectors of plant foods, which consisted of wild squashes, avocados, beans, amaranth and peppers. Between about 3400 and 2300 B.C. the food provided by agriculture rose to 30% and people formed fixed settlements. The cultivation of more 'hybridized' corn and the making of pottery began. Around 200 B.C. the peoples built large irrigation projects and made new food products—from domesticated turkey, tomato, peanut and guava. This brief history of origin of domestication of plants in a remote place in the world reveals that there was continuous human endeavor to increase food production. The motivation for this effort is not the population, which was not so high as of today, and therefore some other factor must have forced them towards such development.

There have been several investigations on the origin of corn. Many investigators are convinced that maize is a descendant of an annual species of teosinte, *Zea mays ssp. Parviglumis*, a native to the Balsas River Valley on the Pacific slopes of Michoacán and Guerrero, in Mexico to a region at elevations between 400 and 1200 meters, while another model associates it with the region of Tehuacán in the state of Puebla, Mexico at elevations between 1,000 to 1,500 meters (see Piperno and Flannery, 2001). But recent studies with the use of accelerator mass spectrometry have fixed the age of maize cobs (*Zea mays L.*) from Guilá Naquitz Cave in Oaxaca, Mexico as 5400 (C^{14}) years B.P. or 6,250 years ago, thus making them the oldest in the Americas. Additional evidence provided by the modern geographical distribution of wild *Zea mays*, and the macrofossil found in the Guilá Naquitz Cave suggested that the cultural practices which caused the domestication of maize occurred elsewhere in Mexico (*see* Piperno and Flannery, 2001; for more details). These authors concluded that until more intensive work is carried out in the region where the teosinte is native, the question of the area where maize was first domesticated would remain in oblivion.

5. DOMESTICATION AND THE HUMAN EFFECTS

Domestication of plants as also the animals especially those which have been used by the humans has taken place along with the evolution of habits of the humans, over long periods of time. There has been mutual dependence between human societies and the plants. Geneticists have been examining the genomes of the domesticated species while the archeologists have been looking for evidence of the human behaviors causing the genetic changes in the course of domestication. Much of the recent studies have been aided with techniques using accelerator mass spectrometer, radiocarbon dating and advanced microscopy. The authors state that domestication is a form of mutualism developing between humans and the plants or animals. It has not happened as an instantaneous change but a long one characterized by changes on both the partners over time and was increasingly interdependent. Early humans intervened in the life cycles of the partners to enable them to increase in their population so that these could enable the humans to grow and expand into new more challenging environments. Zeder *et al.* (2006) have described in detail about these processes of mutual dependence leading to the evolution of modern ways of crop production. It is said that when humans began to harvest, store and plant seeds over long periods, they have created a kind of selective environment to which the plant species have adopted presumably through morphological and genetic changes. There are differences with regard to the 'markers' showing the initial stages of human intervention in the case of root crops, which are propagated, through vegetative parts. Careful examination of the starch grains, for example of the root crops, that were preserved intact on the surface of grinding tools would reveal the nature of domestication (Piperno *et al.*, 2004). The story of evolution of maize and some root crops has been examined using the starch grain analysis of the tools (Dickau, *et al.*, 2007). Crop introductions from one place to another by human intervention have been discussed by Zeder *et al.* (2006). There were trading network linking Africa, Arabia and South Asia which helped in plant introduction to new areas. The banana (*Musa* sp.), a Southeast Asian domesticate, is said to have moved across the 'great lost corridor of mankind' linking east coast of Africa, and the south coast of Asia and introduced into East Africa by three millennia before (De Langhe *et al.*, 1996). Similarly, crops like sorghum (*Sorghum bicolor*), finger millet (*Eleusine coracana*) and pear millet (*Pennisetum glaucum*) considered to have been domesticated in southern margin of the Sahara moved out of Africa moved towards the eastern regions like Oman, and the Indian subcontinent early during 4600 years BP, and these crops were pre-adapted to the Indian summer growing seasons, and helped the rapid growth of Harappan village communities relating to the early Indus valley civilization (Possehl, 1997; Weber, 1991). The origin of bottle gourd and its presence in the New World through the archeological findings has been discussed by Erickson *et al.* (2005) and with radiocarbon dating, this was identified as a domesticated plant in the Americas about 10,000 years ago though it was earlier considered as of African origin.

6. REVOLUTIONS IN AGRICULTURE PAST HISTORY

Progress in agriculture was continuous ever since humans developed the skill to cultivate plants for food. This occurred centuries after centuries, and in different regions either close to, or even far away from, each of the regions. It would be appropriate to make a comparison of how agriculture or domestication of crop plants took place in the Old World. According to Wilhelm G. Solheim II (1972), agriculture was invented at least twice. Plants and animals were domesticated in the Old World and the process was repeated quite independently a few millennia later, in the New World. The agricultural revolution, which was believed to have first occurred about 10,000 years ago in the Neolithic societies of the Middle East, has taken place independently thousands of miles away in Southeast Asia, about 5000 years before. This revolution has involved plants as well as animals not known for the most part in the Middle East. One view is the technological advancement during the period between 13,000 and 4,000 B.C. flourished neither in the Middle East nor the adjacent Mediterranean, but in the northern reaches of mainland of Southeast Asia. Excavations by Solheim and his team revealed that man's first efforts to domesticate the wild plants and animals began here and led to further development of agriculture including horticultural crops in the mountainous terrains of Thailand. He classified the stages of development in the Southeast Asia as: the first period as *Lithic*, when humans used chipped and flaked stone tools; the second, *Lignic*, when wooden tools were used; the third one, the *Crystallitic* period when the domestication of wild plants (about 13,000 B.C) took place somewhere in the northern regions of Southeast Asia, food procurement was based on horticulture; and the fourth one, the *Extensionistic* period, which began around 8,000 B.C. and ended in the early Christian era. The people, who were sedentary, began to travel from the mountain regions to plains. This led to a change in the environment and great progress in farming. The domestication of wild plants and the evolution of agriculture in both the Old World and New World as well as in the Southeast Asian civilization brought in the transition of the people from the hunting to a sedentary habitat. Perhaps with the increase in population, sedentary life which led to an organized living became a necessity, and the animal food brought in through hunting might have become more expensive than food from land.

The agricultural revolution in England took place during 1500 and 1850 (Overton, 2002; BBC, 2007) and more specifically around the year 1750. The farming system prevailing around the 16th century known as 'organic agriculture' was replaced by one, which was characterized by high-energy inputs. Several innovations were introduced in the farming along with it. As the first step of mechanization, a reaper was developed which could harvest and deliver the sheaf. The crops like wheat and barley replaced rye, which was a poor yielder. Similarly crops like clover and turnip were introduced as fodder and were grown in rotation with grain crops. However, turnip did not become popular as a fodder crop until the mid 18th century. The cereal crops were highly productive. The yield of wheat increased over 25% between 1700 and 1800, and then by 50% during the next 50 years. Early in the 19th century the key factor for yield increase was identified as nitrogen. Cereal yields were increased using cattle manure which supplied nitrogen. Though the farmers were aware that the cultivation of legumes like peas and beans, increased soil fertility, they did not know that their roots could 'fix' the atmospheric nitrogen. This new system resulted in dramatic increase in food output. The earlier practice of "organic agriculture" was supplemented with nitrogenous fertilizers, which in turn led to the exploitation of fossil fuels needed for the manufacture of inorganic fertilizers. However Overton was not happy with this change and wrote, "just as a sustainable agriculture had been achieved, the development of chemical fertilizers and other external inputs undermined this sustainability."

Overton (2002) wrote about the British revolution in the BBC, "For many years the agricultural revolution in England was thought to have occurred because of three major changes: the selective breeding of livestock; the removal of common property rights to land; and new systems of cropping, involving turnips and clover. All this was thought to have been due to a group of heroic individuals, who, according to one account, are 'a band of men whose names are, or ought to be, household words with English farmers: Jethro Tull, Lord Townshend, Arthur Young, Bakewell, Coke of Holkham and the Collings". "All these details are in some dispute, but there is general agreement that the role of the 'Great Men' as pioneers and innovators has been exaggerated. 'Turnip' Townshend, for example, was a boy when turnips were first grown on his estate, and he could not, as the textbooks tell us, have introduced them from Hanover. Jethro Tull was something of a crank and not, as we have been told, the first person to invent a seed drill, which in any case was not used by farmers on any scale until a century after his treatise *Horse hoeing husbandry* was first published in 1731".

The 'agricultural revolution' is marked by a significant rise in both land and labour productivity. Why labour productivity rose remains a mystery though it could be related to changes in the deployment of the workforce and the amount of energy expended on the farm, rather than that due to any remarkable technological innovation. But the growing commercialization of agricultural production influenced the emergence of agrarian capitalism (Overton, 2002).

The world's population increased enormously in the last about five decades. The current rate of growth however, is unique in that it is a glaring deviation from the annual growth rates during the most of man's history. According to Freedman and Berelson (1974), the human population during the last millennium increased rather slowly and from the time of about 5000 years ago and until the 17th century, the increase was only 0.1% per year for several decades. It rose subsequently at very high rate, especially in the highly developed European countries and their overseas settlements. In the earlier periods, the high fertility was balanced by high mortality and subsequently the death rates were falling almost everywhere while the birth rate also declined in many nations. Dyson (1999) reviewed the world food trends with respect to population growth. While it is accepted that population increased faster than cereal production since 1984, this has occurred in poorer nations, and the richer nations began reducing grain production which created food shortage. Dyson concludes that in the year 2025, farmers would produce 3 billion tons of cereals to feed around 8 billion people, and yield from one hectare would have to be around 4 metric tons. The yield increase has to be obtained by the application of our present knowledge and technologies and essentially the farmers have to inevitably use more of synthetic nitrogen fertilizers which have to be nearly double that as used earlier. It is highly unlikely that there will be any significant breakthrough that will help raise world food production. Even though food production would continue to rise, the rapid population growth will outweigh the food supply, which will eventually lead to expansion of world food trade to meet the shortage.

There have been efforts to increase food production too, using modern technology. This is a deviation from Dyson's views. There are some new problems as is seen now. The cost of production increased though, while the purchasing power declined. Producing more food necessitated greater use of fertilizers, the cost of which went up over the last few decades, and is still on the increase. Food, famine and fertilizers are inter-related. Soil, water and air are important as inputs for food production. Fertilizer is required to supplement the soil fertility as also for enhancing the crop production. If food production is insufficient or grossly inadequate, famine invades, as it occurred several times in the human history.

7. BREAKTHROUGHS IN RICE AND WHEAT

The year 2007 is the 100th anniversary of the American Society of Agronomy, and the society has given recognition to two of its members (Milford and

Runge, 2007). One is Henry M. Beachell, internationally known and most long-lived 73-yr member, and the other, Norman E. Borlaug well known for his accomplishments and both were recognized for ushering in "The Green Revolution". Beachell was responsible for the revolution in rice production, while Borlaug was with wheat. Both attributed their success to the discovery of the "dwarf gene" in both the plants. The dwarf gene for wheat was brought from Japan to Washington State by another agronomist Orville Vogel, and rice from Taiwan to International Rice Research Institute. They recognized the great potential of the dwarf species since these were less susceptible to lodging under heavy fertilization. They developed new high yielding varieties in collaboration with many scientists and their efforts made a major impact for increasing rice and wheat production to unprecedented levels. As a result, countries like India, Pakistan, Indonesia, the Philippines which were poor producers of grain for their people in the 1950s became self-sufficient by 1970s. (Beachell died on 13 December 2006, 3 months after his 100th birthday).

8. MARK A. CARLTON AND THE EARLY WHEAT REVOLUTION

Carlton was the first president of the American society of agronomy. Paulsen (2001) has described the achievements of Carleton in his article "The Agronomic Legacy of Mark A. Carleton". The author has given details of how the ASA was formed in his article. "A small group of persons, including Carleton, met at the University of Chicago on 31 Dec. 1907, to form the American Society of Agronomy.

Carleton had conducted much of the correspondence that led to organization of the Society and was unanimously selected as its first president. The early growth and development of the Society were greatly influenced by him".

Carleton was responsible for the development of the hard red winter wheat (*Triticum aestivum* L) and *Triticum turgidum L. var. durum*, widely known as *durum* wheat on which the wheat industries in USA heavily depended. He introduced wheat cultivars which were resistant to diseases. His introduction includes the hard winter wheat which became the most popular one in the central and southern Great Plains, *Kharkof* wheat which became an important variety grown in the central and northern plains, the Crimean wheat which was a parent of many later cultivars., and *Kubanka* which was much needed for wheat industry in the northern Great Plains. "He was considered the most respected agronomist in the USA and he defined the profession in principle and in practice." Such was the greatness of this agronomist who revolutionized wheat cultivation and who was dynamic in traveling and searching for the traits of several wheat cultivars for their suitability to be introduced to USA. But the later part of his life was full of sufferings, and is also described by Paulsen. Paul de Kruif (1928) has mentioned about Carleton whom he calls "*the wheat dreamer*" in his book "Hunger Fighters."

There has been no record of men who fought human hunger and the sufferings they underwent in the course of such "fights". One such person is Mark Alfred Carleton who brought the magnificent wheat *"Kubanka"* from its original home on the

Turghai steppe of western Asia and was cultivated nearly four millions of acres in the American Northwest. Carleton discovered that the lands in America resembled the dark fallow land of Russia and was successful in introducing it in USA. In 1917 from within his own office, charges were preferred against Carleton by his known men and the charges were trivial. On the 26th April, 1925, at Paita in Peru, Mark Alfred Carleton died, aged fifty-nine of acute malaria and with a broken heart. Kruif further laments that there were nothing like a monument for this man who was 'child of Nature'. But the great fields of Kubanka could bring back the memory of this great man.

William Jasper Spillman another agricultural scientist born in Missouri (October 18, 1863 - July 11,1931) is considered to be the founding father of *agricultural economics*. His early days were spent in his ancestral farm with a large family dependent on him since his father's accidental death in 1871. He enrolled at the University of Missouri for B.S. and then M.S. in 1890, later joined the faculty at Pullman. Here he made the scientific discovery independently the 'Law of Heredity', and it is said that, had he not been preceded by Mendel forty five years earlier, Wilman would have his name known throughout the world. His discovery of the law of inheritance was made while carrying out wheat hybridization work. Laurie Carlson (2005) wrote about Spillman, "He remained linked to his Missouri roots by participating in the National Grange his entire adult life, while he also was a member of the American Academy of Science and received an honorary doctorate for his pioneering work in genetics". His work on plant genetics resulted in agricultural overproduction and causing depopulation of rural America in the 1930s. He formulated a national agricultural allotment program, which later became the foundation for *"federal farm programs"*. He published two books in 1924 and 1927, namely "The Law of Diminishing Returns", and "Balancing the Farm Output: A Statement of the Present Deplorable Conditions of Farming, its Causes and Suggested Remedies" which helped lay down future agricultural policies in USA. He was a teacher, educator and an agriculturist who developed scientific methodology for solving the agricultural problems "To the farmers, he was not the government's expert, but a practical man who knew what he was talking about and, as well, knew when to listen. Farmers said, 'Don't send me no experts, send Spillman'(Information derived from Wikipedia, the free encyclopedia News July 6, 2007). In honor of him a memorial was raised in the Washington State University campus. "Stone honoring wheat scientist planted again in a place of honor on the Washington State University campus" news appeared in Seattle Times, October 24, 2006.

9. MYTHS ABOUT THE HUNGER AND FOOD PRODUCTION

The Worldwatch Institute has discussed the prevailing twelve myths about the world shortage of food, in its summer issue of 1998, volume 5. No.3. It is very often said there is not enough food produced. It is not true and the fact is though food is abundant, the purchasing power of the poor is very much less. Altieri of the Food First Research Institute has explained why the AGRA (Alliance for a green revolution in Africa) will not solve the problem of hunger and poverty in Africa, in the Policy Brief No. 12, October 2006. The Rockefeller Foundation and the Bill and Melinda Gates Foundations have jointly gifted huge amounts for a 'green revolution' in Africa. This proposal has received severe criticisms on the one ground that it has not taken into account the failures of the first Green Revolution and the causes for that. Holt-Gimenez *et al.* (2006), of Food First/ Institute for Food and Development Policy, have stated that the problem is not the production of food, but the poor people in these countries cannot afford to buy all the food that is produced in their countries. He has also pointed out that even in countries where the green revolution was claimed as success do have problems now after nearly 40 years. There have been soaring suicide rates in the Indian farmers in Punjab and the reason is their fertilizer debts have been too heavy, while at the same time their soil productivity has diminished.

10. FOOD PRODUCTION IN CHINA

Sylvan H. Wittwer (1987) former Director of Agricultural Experiment Station, East Lansing, Michigan made an in-depth study of the development of agriculture in China and made five trips during 1980 to 1987 and presented along with three Chinese authors a detailed report in 462 pages' book. China is able to feed over a billion people or about 22% of world's population on 7% of the earth's arable land, as at the time of the report, and it has set the hallmark of success in food production. Within the period from 1978 to 1986, it has changed over from a state-controlled economy to a market-oriented one with economic incentives for increasing production. Wittwer attributes three factors which influenced their production. The first is labeled *economization*, which relates to marketing and market economics which assumed a dimension as never before. Second is the emphasis on *diversification*. Farmers are motivated to grow or produce that which is most profitable to them. A third area of transition is the *modernization*, which includes exploitation of scientific and enlightened farming systems.

Writing about the food production in China, Lester R. Brown (1997) mentions that raising wheat yields is increasingly difficult, as aquifers are getting depleted, and also the response to additional fertilizers is on the decrease. Furthermore the country's explosively growing cities divert irrigation water away from agriculture. It is seen that the Yellow River is now running dry for several weeks each spring, and progressively longer periods year after year, which makes the farmers downstream face shrinking irrigation water for crops. For some weeks during the spring and summer year of 1996, the river ran dry and the coastal Shangdong Province which produce one fifth of the China's wheat did not get the irrigation water from the river.

11. THE ROLE OF BIOTECHNOLOGY

Halweil (2000) of the Worldwatch Institute talks about the use of agricultural biotechnology in relation to food and agriculture. He considers that the principal causes of poverty and hunger are largely based on the socioeconomic systems and not necessarily due to lack of adequate technology, and there is no technology that immediately can save people from hunger and poverty. His second concern is the fact that the global biotechnology industry has invested enormously on the small range of products, which have "large and secured markets within the capital-intensive production systems of the First World—products which are of little relevance to the needs of the world's hungry". Many crops from apples to lettuce to wheat, have been genetically engineered and are under commercialization. Of the transgenic crops developed as at present, soybeans and corn occupy about 54 .and 28% respectively of the global transgenic area. Some of the transgenic crops are developed for resistance to spraying of herbicides and insecticides and these traits help large-scale industrial farmers to reduce the costs of crop production. However, there does not appear to help the farmers of the developing nations and the needs of the hungry. Halweil cites the joint report by the National Academy of Sciences and seven other academies around the world which concluded that transgenic plants are not being used in many of the developing world where the need for such crops is the greatest. He has expressed a genuine concern about biotechnology. The privatization for example, of the germplasm puts public sector agricultural research at a disadvantage, and would threaten the majority of small farmers in Africa, Latin America and Asia, who depend on the seeds saved from their own crops. However, he stresses the importance of biotechnology to develop varieties responsive to low levels of soil fertility, and those tolerant to salinity and drought, and to low agrochemical inputs. In this connection it is important to note that there are several considerations which do not favour genetically engineered crops.

Soil salinity and drought have very detrimental effects on the crops. Flowers (2006) invited speakers from Australia, Europe and USA to contribute papers on 'salinity' in the meeting held in Barcelona. Salinity has threatened many parts of many countries and with increasing population it has become necessary to tackle the problem and make these lands more cultivable. The editor of the meeting states, "More worryingly, agriculture has itself also brought salt to the soil; both through irrigation and forest clearance. The combination of population growth and land degradation has led plant scientists to the view that increasing the salt tolerance of crops will be an important component of future agriculture."

Many of the discussions have been published by the journal of experimental botany in 2006 as special issue. Some of these are included here. Munns *et al.* (2006) have reviewed the physiological mechanisms of salt-tolerance. Identification of new genetic sources having salt tolerance is emphasized. Screening for the salt tolerance traits in wheat in particular, and the future line of work are indicated. The physiological traits relating to salt tolerance are 'mechanism for Na extrusion' and sequestration of Na and Cl in the vacuoles of root and leaf cells, to cite a few. Cuartero *et al.* (2006) have proposed three techniques for introducing salt tolerance in tomato: 1. Treatment of seedlings with drought or NaCl, 2. Mist sprayed to tomato plants to improve vegetative growth and yield in saline conditions, and 3. grafting tomato cultivars onto rootstocks having salt tolerance. Colmer *et al.* (2006) have proposed that the genetic information on salt tolerance of selected species in this tribe of grasses be pooled, and the potential ones of the wild species used by geneticists to improve salt tolerance in wheat.

Attempts to improve crop production in the saline soils through plant breeding have not been very successful, primarily because of the multigenic control of the adaptive mechanism. The transfer of genes through genetic engineering is a likely successful approach. Barkla *et al.* (1999) state that the technique of gene transfer should involve the tonoplast Na+/K+ antiport, compatible solute synthesis and regulation of water channel activity and expression. Some of these include enzymes involved in the synthesis of the compatible solutes and those for the water-conserving metabolic adaptations. So far, only the mechanisms are understood, but it is a long way to actually carry out the gene transfer. A recent report in the ENS-news august 2001, states that plant biologists Blumwald and Hong-Xia Zhang at the University of California and University of Toronto have developed the first truly salt tolerant crop-a genetically engineered tomato plant that grows in salty irrigation water. The scientists have also shown that these tomato plants produce higher levels of a naturally occurring protein, a transport protein and that the gene responsible for the production of this protein was isolated from *Arabidopsis*. However, a poll conducted in USA showed that more than 50% of respondents do not consider such foods as safe to eat.

A genetically engineered variety of potato, which will be rich in protein, has been developed at the National Center for Plant Genome Research at New Delhi, and is expected to have 14% protein content. Normally potato contains very large amounts of starch. By the addition of protein in potato genome, it will supply better nutrition for those who include potato in their diet because of its cheapness. It is expected to be released to the farmers soon, says a report dated December 8, 2000. There has been large scale opposition for growing the genetically modified cotton crop in Gujarat. The Gujarat government issued an order that the crops of Gujarati farmers in Gandhinagar district using a variety of cotton seed known as "Navbharat 151" would be burnt on the grounds that it contained the forbidden genes. This is under a mistaken impression that it contained assumed mythical "Terminator gene" which is said to be implanted in the new cotton varieties in order to stop their spread. Due to the great delays by the central government in approval of the cotton variety, the Bt cotton is believed to be distributed surreptitiously.

It is important to note that the relatively better-off farmers largely utilize the technological benefits of the Green Revolution. The majority of the farmers of the world's hungry are bypassed, or marginalized by the Green Revolution packages. A report in the early 2000 describes a situation that has arisen in Indian agriculture. It is anomalous that when the economy shows an inflation of around 7.5%, the prices of agricultural commodities overall show around only 1% rise between April and October 2000. It is also reported that open market wheat and rice prices have declined by 1.8 and 7.6%, and even those of millets like jowar, bajra, ragi and maize, have declined very sharply. Thus, the recovery price for farm produce has not been commensurate with the increasing input costs. The latter however varies in different states, because many states have zero power tariffs and low water charges for farmers. The opposition charge against the then National Democratic Alliance government was that the farmers interest was not protected, by way of giving support price or buying the commodity from the farmers at a reasonably attractive prices and debit it to the taxpayers' account. But even with the change in the government, the situation regarding the support price for the farmers especially wheat has remained the same, i.e., it is not commensurate with the cost of producing the grain. There has developed an erroneous approach for bringing down the price when the grain production has fallen below the expectation, to import for example, wheat, and at a high cost. The farmers clamor that if support price was increased, they would be able to increase the production. It is interesting that what began in the 1960s as a buffer stock policy for stabilization of agricultural prices has become now a Frankenstein monster. Such a politically driven policy will only negate the basic agricultural philosophy which lie in rational crop selection, food security, enhanced productivity and effective management of water resources and transforming agriculture from a traditional occupation into a more powerful and decisive factor in the nation's economy. Despite these facts like the fall in grain prices, larger sector of the poor do not have the buying capacity for food. There is one reason that many of these go in for attractive consumer goods at the expense of food. Due to large scale manufacture or liberalization policy, consumer goods have flooded the markets, and induce the poor to go in for these. School going children in the poorer sector prefer to take to small jobs to supplement the family income, rather than go to school.

Food procurement at minimum support prices, has reached all time high, estimated at 62 million tonnes, while the expected storage capacity is only around 30 million. The food stock is likely to exceed 75 million tonnes by early 2002. The policy of minimum support price of the government originally intended to protect farmers against the likely farm price crashes has largely benefited the well-to-do farmers in Punjab, Haryana and Andhra Pradesh which constitute the 80% procurement of food grains in the country. Now there are two questions, whether the govt policy of support price could be continued at the expense of the tax-payers money, and whether the food storage capacity should be increased for the benefit of the population remaining below the poverty line. However, the fact that food production can be increased by constant efforts of the intelligent farmers, but at what cost on the soil and water quality and environmental safety.

12. AGRICULTURE: REDEFINING THE OBJECTIVES IN USA

Agricultural Research and Extension in US—— The need for Redefining Their Objectives: Describing about the world agricultural production and the U.S. farm dilemma, Avery (1985) states that the predictions made in the Global 2000 Report presented to the President Carter in 1980 did not fully come true. It was estimated that the world food demand would increase greatly in the following 20 years, which is around the year 2000 and that the developed countries would have the responsibility to produce more food. Avery states that after 5 years of the prediction the American farmers faced problems of mounting food surpluses, with the increasing farm debt and farm subsidy costs. The demand for US farm products decreased and forced the government to revise the farm policy. While these are the situations in the US, there are famines reported all over the world, especially in the African countries where there were thousands of deaths due to malnutrition in particular. However, the performance in the food production sector in the developing countries is commendable. World food output rose by 25% between 1972 and 1982 a period of ten years, which was an all-time high. This improvement is due to implementation of the technological reforms. Fertilizer use doubled in these countries and its production tripled. The farm and food policies of these developing countries have improved for the better, focusing on productivity. Thus the constraints on food production expected in the Global Report turned less severe than predicted. However, this situation does not mean the efforts to increase food production and to develop new technologies should come to a halt.

The increasing population all over the world does not keep one rest in complacency. In fact, there is constant threat from famine and undernourishment of many of the people in the developing and even the developed nations. The threat of famine and maldistribution which force developed countries to supply the needy in emergency is imminent, as we boast over our achievements. The agriculture secretary of the US,

Dan Glickman announced the supply of 350,000 metric tons of wheat, corn, rice and other commodities to African countries, in the year 2000, and he said, "We are targeting these donations to drought-ravaged or war-afflicted countries, especially in the Horn of Africa where the drought continues to widen. These donations reflect the American tradition of sharing our abundance with the hungry, the displaced, and those who most desperately need our help around the world".

Avery concludes on the future of the US farm policy thus. "The US farm policy of the future must be geared to competing for buyers who have more alternative sources of supply than ever—their own agricultures all over the globe, and more synthetics and subsidies. This means that our policies must be designed to reduce costs per unit and to provide farmers with the latest technology.....Researchers need to look at farmland not only in the traditional sense but also as a potential source of biomass and the various kinds of complex chemical feedstock that could be produced from genetically engineered plant life. One thing seems certain: the price supports, land diversion, and storage programs that have dominated US farm policy for the past 50 years work against in US farmer in a world of high technology and rising productivity".

J.B. Kendrick, Jr. (1986) narrates his experiences on his association with California Agriculture. According to him the productive capacity of US agriculture is one of the highest in the world, and because of this, US produces food and fiber far in excess of their domestic demands.

Therefore for economic survival, their major crops and specialty commodities should find rewarding foreign markets. Furthermore, the productive capacity will be increased greatly with the incorporation of "space age" technologies and application of biotechnological research. He predicts that small and moderate-size farms would require off-farm income and assistance because the large scale farms would by the year 2000 market 75% of the farm products utilizing 60% of the total US farmland. This change will influence the US public policy relating to the agricultural research and extension. He further states that "International trends will also affect the structure of US agriculture and influence publicly supported agricultural research and extension. The rapid increase in productivity that characterized US agriculture in the 1950s and '60s is now occurring in all the major agricultural regions of the world......Rising worldwide productivity means that US agriculture will face fierce competition for international markets......For US agriculture to remain competitive in this international trade environment, it must continually improve its productivity and its efficiency. This almost guarantees the continuation of research into biotechnological applications for agriculture and into further development of systems for managing information vital to successful agricultural activities.......The most significant change required for the future involves Cooperative Extension.....It must assist these farming units in adapting the new technologies....and it must help the moderate and small-unit farmers' market efficiently and effectively at home and abroad".

When LGCA were established 140 years ago, agriculture was the main occupation of the people. There have been many changes in the land grant colleges of agriculture (LGCA) which had a traditional agriculture base and now the technology is much advanced. The people employed in the information based society are overwhelming. The agricultural colleges which are now known as Land Grant Universities with several departments other than agriculture, have been receiving enormous funding and grants. They encourage *innovation*, and provide strong base and leadership in research in biotechnology and genetic engineering. However, there is continuing need to encourage students to become strongly science-based, along with their broadening of the information base.

13. GLOBAL FARM CRISIS

The editor Ed Ayres (2000) writes that as globalization accelerates, the food industry takes over many of the more profitable functions once carried out by the farmers. In the United States, the consumer's food dollar actually going to the farmer has declined from more than 40 cents before 1950 to about 7 cents today about 5 cents of the dollar goes for the wheat, and another 5 cents goes for the wrapper and a large percentage goes for the marketing. It is interesting to note that a big change on the farming front is taking place. In countries where corporate farming has grown, many of the farmers are out of farming. In Nebraska and Iowa, the US breadbasket states, the farm population is reduced and more will be driven out in the next 2 years. This trend will be observed in Sweden and Poland and in the Philippines too. As small farms are taken over by larger ones, the rationale often given is that the large enterprises are more efficient or productive. But this is in large part a myth, says Halweil. While a large monoculture operation may produce more output per acre of that crop, the small farm involved in traditional polyculture, farming is more intensive and efficiently use the resources with the result significantly more food per acre could be produced.

14. FOOD PRODUCTION AND GLOBALIZATION

Food production is to be increased to meet the larger demand of the future population. Marketing of food and food products is another important of aspect which involves food harvesting, packaging and marketing. The developed countries have surplus food production and have a well-developed marketing systems. These are also principal exporters of food grains and fruits and food products to the developing countries which are still in deficit. Recently this export promotion has been encouraged through globalization. The exchange of food across several food producing nations and those which are in deficit, is said to be prevalent over centuries although the study of this in relation to anthropology is relatively new according to Phillips (2006) who has discussed many aspects of food and globalization recently. The earliest examination of food globalization was by Mintz (1985). Several reports on globalization appeared subsequently. It is said that a new era on the global food regulation is just begun indicated by a flexible food production system, and formation of corporate bodies to operate international trade with free movement of goods. This era is called a "quiet revolution" (Schertz and Daft, 1994) and is also considered a threat to national market development. However this international trade has helped in the adoption of agricultural practices like planting, picking and packaging. Some economists have considered this in the light of a shift towards a flexible labor relation for food production and export and Barnet and Cavanagh (1996) call this as "feminization" of labor. Women workers are employed in the food industry in the international corporations as packers, food processors, supermarket cashiers and providers of food service (Barndt, 1999). The benefits or otherwise have been discussed by several economists (Ritzer, 1993, Watson and Caldwell, 2005). Phillips (2006) considers that the introduction of food-related transnational corporations into the developing countries is likely to produce negative effects on nutrition of the people since it replaces or modifies the diet of the local people and there is evidence that obesity has increased as a result of imported foods (Evans *et al.*, 2003). However it is true that the new supermarkets in the developing countries offered greater dietary choice, but it is noted that this affected small farmers. The study of globalization of food has led to the view that many multilateral financial lending institutions like the World Bank, International Monetary Fund, international trade agreements are 'handmaidens' of transnational corporations to promote 'new global order' (McMichael, 1999). Phillips has discussed many issues relating to food and globalization and mentions about the role of science and technology and the farmers for adopting to global markets. This is important especially with the introduction of genetically modified crops which would likely feed the world population by the year 2030. While this development on agriculture and marketing of food products are likely influence the developed countries which also directly affect or influence the farming patterns, the developing countries have to be cautious about these transnational corporations.

REFERENCES

Araus, J.L., J.P. Ferrio, R. Buxó and J. Voltas. 2007.The historical perspective of dryland agriculture: lessons learned from 10,000 years of wheat cultivation. J Expt Bot 58:131-135.

Avery, D. 1985. U.S. Farm Dilemma: The global bad news is wrong. Science 230:408-412.

Ayres, Ed., Worldwatch News Release September 1, 2000. Worldwatch Institute, 1776 Massachusetts Ave., NW. Wash DC 20036.

Balter, M. 1999. 'A long Season Puts Catalböyük in context'. Science, 286:890-891 (*see* Wirtz and Lemmen, 2003).

Barkla, B.J., R. Vera-Estrella and O. Pantoja. 1999. Towards the production of salt-tolerant crops. Chemicals *via* Higher Plant Bioengineering, edited by Shahidi *et al.* Kluwer Academic/Plenum Publishers, NY, *pp* 77-89.

Barndt. D. ed. 1999. *Women Working the NAFTA Food Chain: Women, Food & Globalization.* Toronto: Second Story Press. (*see* Phillips, 2006).

Barnet, R. and J. Cavanagh. 1996. *Global Dreams: Imperial Corporations and the New World Order.* New York: Simon & Shuster. (*see* Phillips, 2006).

Boserup, E. 1981. Population and Technological Change. University of Chicago Press (*see* Wirtz and Lemmen, 2003).

British Broadcasting Corporation. 2007. http://www.bbc.co.uk/history/british/empire_seapower/agricultural_revolution_01.shtml.

Brown, L.R., 1997. Facing the challenge of food scarcity: Can we raise grain yields fast enough? In Plant Nutrition-for sustainable food production and environment, , Developments in Plant Sciences, Vol 78, pp 15-24.

Carlson, L. 2005, Forging His Own Path: William Jasper Spillman and Progressive Era Breeding and Genetics. Agricultural History79, Number 1

Colmer, T.D., T.J. Flowers and R. Munns. 2006. Use of wild relatives to improve salt tolerance in wheat. J Expt Bot 57:1059-1078.

Cuartero, J., M.C. Bolarin, M.J.Asins, and V. Moreno. 2006. J Expt Bot 57:1045-1058.

Dickau, R., A.J. Ranere and R.G. Cooke. 2007. Starch grain evidence for the preceramic dispersals of maize and root crops into tropical dry and humid forests of Panama. Proc Natl Acad Sci 104:3651-3656.

Dyson, T. 1999. World food trends and prospects to 2025. Proc Natl Acad Sci USA 96:5929-5936.

Erickson, D.L. et al. 2005. An Asian Origin for a 10,000 year old domesticated plant in the Americas. Proc Natl Acad Sci USA 102:18315-18320. (*see* Zeder *et al.*, 2006).

Evans, M. R. Sinclair, C. Fusimalohi, V. Laiva'a and M. Freeman. 2003. Consumption of traditional versus imported food in Tonga: implications for programs designed to reduce diet-related non-communicable diseases in developing countries. Ecol Food Nutr. 42:153-176. (*see* Phillips, 2006).

Flannery, K.V. 1973. The origins of agriculture. Ann rev anthropol 2:271-310.

Flannery, K. V. and J. Marcus. 2001. Richard Stockton MacNeish: April 29, 1918-January 16, 2001. Washington, D.C.: National Academy Press.

Flowers, T. 2006. Preface. Plants and Salinity. Special issue. J Expt Bot 57: iv

Freedman, R. and B. Berelson, The Human Population, Scientific American, 1974, 231:31-39.

Glickman, Dan, US Agriculture Secretary, Release No. 0238.00 by Andy Solomon, July 17, 2000.

Halweil, B. 2000. Testimony from Senate Subcommittee on International Economic Policy, Export and Trade Promotion at the Hearing on "The Role of Biotechnology in Combating Poverty and Hunger in Developing Nations", July 12, 2000.

Holt-Gimenez, E., M.A. Altieri and P. Rosset. 2006. Ten reasons why the Rockefeller and the Bill and Melinda Gates Foundations' Alliance for Another Green Revolution will not solve Problems of Poverty and Hunger in Sub-Saharan Africa. Food First Policy Brief No.12. October 2006.

Kendrick, J. B. Jr., Reflections and projections (Calif Agri 40:2-3, 1986, May-June, nos.5 &6,).

de Kruif, P. 1928. *The wheat dreamer Carleton. In* Hunger fighters. Pp 2-30

Harcourt, Brace & World, New York, NY

de Langhe, E. *et al.* 1996. Plantains in the early Bantu world. In *The Growth of Farming Communities in Africa from the Equator Southwards* (Sutton, J.E.G., ed.) *Azania* 29-30, pp. 147-160, The British Institute in Eastern Africa. (*see* Zeder *et al.*,2006).

MacNeish, Richard S., 1964. The Origins of New World Civilization, Scientific Amer 211:29-37.

McMichael, P. 1999. Virtual capitalism and agri-food restructuring. In *Restructuring Global and Regional*

Agricultures, ed. D. Burch, J. Goss, G. Lawrence, pp. 3-22. Aldershot, UK: Ashgate. (*see* Phillips, 2006).

Milford, M.H. and E.C.A. Runge. 2007. Beachell and Borlaug, two giants of the American society of agronomy's first century. Agron J 99:595-598.

Mintz. 1985. *Sweetness and Power: The Place of Sugar in Modern History.* New York: Penguin (*see* Phillips, 2006).

Mithen, S. 1996. The prehistory of the mind:A search for the origins of Art, Religion and Science. London:Thames and Hudson. (*see* Weisdorf 2005).

Munns, R., R A. James and A. Läuchli. 2006. Approaches to increasing the salt tolerance of wheat and other cereals. J Expt Bot: 57:1025-1043

Múzquiz, J. L. L. 2000. "The Cañün de Diablo. A window to ancient history in the northeast of Mexico. México desconocido # 277 / March 2000.

Overton, M. 2007. Agricultural Revolution in England 1500 - 1850. BBC Home Page History, 20 May 2007.

Paulsen, G.M. 2001. The Agronomic Legacy of Mark A. Carleton. J. Nat. Resour. Life Sci. Educ. 30:120.123.

Phillips, Lynne. 2006. Food and Globalization. Annu Rev Anthropol 35:37-57.

Pickersgill, B. and A.H. Bunting. 1969. Cultivated plants and the Kon-Tiki theory. Nature 222:225-227. (*see* Flannery, 1973).

Piperno, D.R. and K.V. Flannery. 2001. The earliest archaeological maize (Zea mays L.) from highland Mexico: New accelerator mass spectrometry dates and their implications. Proc National Acad Sci 98:2101-2103.

Piperno, D.R. *et al.* 2004. Processing of wild cereal grains in the Upper Paleolithic revealed by starch grain analysis. Nature 407:894-897. (*see* Zeder *et al.*, 2006).

Possehl, G.L. 1997. Sea faring merchants of Meuhha. In *South Asian Archaeology* (Allehin, B. ed.), pp. 87-100, Oxford & IBH Publishing Co. (*see* Zeder *et al.*, 2006).

Ritzer, G. 1993. *The McDonaldization of Society.* Thousand Oaks, CA: Pine Forge Press. (*see* Phillips, 2006).

Rymer, Eric. 2006. Farming in early India. Notes on the development of farming in ancient India. http://www.historylink101.com/lessons/farm-city/india-city.htm

Schertiz, L.P. and L.M.Daft. *ed.* 1994. *The Food and Agriculture Markets: The Quiet Revolution.* Washingotn DC. Econ Res Serv. U.S.D.A. Dept of Agric, Food and Agric Comm. Natl. Plan. Assoc. (*see* Phillips, 2006).

Smith, B.D. 1997. The initial domestication of *Cucurbita pepo* in the Americas 10,000 years ago. Science 276:932-934.

Smith, B.D. 1998. The Emergence of Agriculture. Freeman and Co. Publishers, New York (*see* Wirtz and Lemmen, 2003).

Solheim II, Wilhelm G., An Earlier Agricultural Revolution, Scientific Amer. 1972, 226:34-41.

Wadley G, and A. Martin. 2000. The origins of agriculture: a biological perspective and a new hypothesis. Journal of Australasian College of Nutritional and environmental Medicine 19, 3–12. 1165

Watson, J.L. and M.I. Caldwell. Eds. 2005. *The Cultural Politics and Eating: A Reader.* Oxford, UK: Blackwell. (*see* Phillips, 2006).

Weisdorf, J.L. 2005. From foraging to farming: explaining the Neolithic revolution. J econ surveys 19: 561-586.

Wirtz, K.W. and C. Lemmen. 2003. A global dynamic model for the Neolithic transition. Climatic Change 59:333-367.

Wittwer, S. H., Feeding a Billion—Frontiers of Chinese Agriculture. Michigan State University Press 1987.

Zeder, M.A., E. Emshwiller, B.D. Smith and D.G. Bradley. 2006. Documenting domestication: the intersection of genetics and archaeology. Review. Trends in Genetics 22:139-155.

CHAPTER-II

FAMINE

"If government knew how, I should like to see it check, not multiply the population".

Ralph Waldo Emerson

1. FAMINE: GENERAL

W.A. Dando (1983) presented a detailed study on famines and stated that famines were occurring even prior to biblical times. These were unexpected, and irregular causing cultural hazards and death of many people. Ten famine periods within two millennia were recorded, leading to social disruption and mass death in 'The Promised Land', likely due to cultural decisions and indifferences. Dando stated that the first famine took place in 1850 B.C. while the last one was in A.D. 46 as shown in the records of the Old and New Testaments. There were 95 famines in the British Isles, and 75 in France during the periods from 501 and 1500 A.D., 150 in Eastern European countries between 1501 to 1700 A.D., 70 in India during 297 to 1943 A.D., 1829 famines in China between 108 B.C. and 29 A.D. In Russia, there occurred 77 famines between 971 and 1970 A.D.

Famines which occurred in major urban centres were a fore-warning for the twenty-first century. The major factor causing famine and aggravating the food problem is the increase of human population. It is estimated that the world metropolitan population will exceed six billion by the beginning of the 21st century and the urban food shortages would be dangerously high. Furthermore, unmanaged urbanization leads to social discontent, crime, environmental pollution causing health hazards, inefficiency and graft in government, and social unrest. According to Dando, stagnation in the income of the poor on one hand, with the enrichment of a small but powerful segment of the elites on the other, have led to a very slow increase in calories consumed per person and a faster growth in nonfood and non-agricultural goods. Furthermore, wrong policy decisions and investments within the developing nations would invariably result in an increase in hunger along with an increase of economic growth. Famine also reflects the manifestation of an imbalance between the total available food and population. If it occurred in a restricted geographical area it leads to the spread of diseases and death from starvation in that region. Natural hazards like drought, floods, excessive heat, and storm, invariably result in crop failure followed by famine. Cultural hazards arise through a wrong decision by someone or organization political or otherwise, and could be against those who are in dire need resulting in mass starvation. It includes war, riot, and revolution of many types.

Dando refers to the Bible as an excellent source of famine data especially those which occurred in the Middle East, surrounding the Mediterranean Sea and the high Trans-Jordan Plateau, the mountains of Lebanon and the southern deserts. This land bridge connects the Nile Valley, Mesopotamia, the kingdoms of the northern Mesopotomia with their productive irrigated lands, their prosperous communities and advanced civilization. There were fights between the nations, viz., Egypt, Mesopotomia, Asyria, Greece and Rome, for the control of this bridge. The climate of the 'Promised Land' at that time, ranged from wet winters with torrential rains and dry summer which extended for five months. There were constant quest for food and the cultural survival by the inhabitants. In the first chapter of the Bible, food had a definite logical place in the Creator's plan: "I give you all the seed bearing plants that are upon the whole earth, and all the trees with seed bearing fruit; this shall be your food" and the basic diet of the early Hebrews were like those of the higher mammals. The three agricultural products that were known in the early era of agriculture were "Wine to make them cheerful, oil to make them happy, and bread to make them strong". Bread was the staple food made from wheat or barley flour. A typical diet was solely from the plant products and eating meat was not permitted. Later the new food scheme, i.e., *Second Food Scheme*, was formulated which allowed meat consumption, provided there is no 'blood'. And then there was a *Third Food Scheme* which segregated the Hebrews from other peoples. The final and Fourth Food Scheme was promulgated by Jesus who declared "all foods are clean" and that it was not what a person ate that made a person unclean, but what he thought, spoke and did. This scheme enabled the inhabitants of the 'Promised

Land' to live among the people of the world. Dando concluded that rapid urbanization, inadequate international emergency food distribution networks, cultural bias, and human indifference combined with inefficient, improper land use caused stress on governments in their efforts to provide food for the increasing urban-metropolitan population and these areas were vulnerable to famine.

2. FAMINE AND HUMAN POPULATION

The human life is a small speck on earth and if we look at it as a small globe two feet in diameter, most of life would be contained within the surface and the habitat of humans would only be a thin layer of the earth's surface. Looking back to about 10,000 years, the population which was only a small gathering of 5 to 10 million humans in the Neolithic age,. was not enough to influence on the ecosystem then prevailing, and this situation continued for the next 10,000 years. Only in the last few decades, man as a species brought about changes, unprecedented before during several millennia. It is appalling to realize that forests that grew over centuries and soils that took millions of years to develop are now being used up to such an extent as to threaten human survival itself. Looking at the demographic figures, the population of the world at mid-century was 2.5 billion and it passed to five billion in the 1980s. According to the projections of the United Nations Population Division, the next 35 years i.e., up to year 2025, will see an increase to 8.5 billion (Keyfitz, 1989). The question-is whether we should worry about the absolute increase of 3.2 billion people or we should be complacent with the fact that the rate of increase is slowing. The following facts are revealing. The total population increased by 9% between 1980 and 1985 and the projected increase between 2020 and 2025 will be only 4%. However the absolute number of births would come down even to the high levels of the 1990s, while the population curve will continue to go upward beyond the second quarter of the 21st century. In other words, the hungry will represent a declining fraction of the total population while the absolute number of the hungry will be on the increase.

Some people do not accept the fact that increasing population implies increasing mouths to feed. As far as India is concerned, all the population control programs seem to begin with one basic premise: the population bomb is out of control. If the population is controlled, the lives of other people may improve and the country may proceed on the path of development. On the other hand, the rapid population growth in India is a symptom rather than the cause of arrested development and stunted lives. Villagers in India generally view children as assets because they help both at home and in the fields from their early ages. It is especially true that fishermen communities need more hands to help their fishing profession. Furthermore, instead of emptying the family's food-bin and water resources, they supplement it, even during the periods of drought. Because the State has failed to guarantee the poor, adequate nutrition or access to cheap healthcare, many children die prematurely and so more children are produced to ensure help and security in the old age. Population explosion is revealed by the following demographic statistics. In 1900, the global population was 1.7 billion and between the years 2150 and 2200, it is expected to reach 11.6 billion. Looking at the figures of population growth in India, it was 3.2 million in 1947, and is now over a billion. Mrinal Pande, a freelance journalist in Delhi, considers that there is enough food, but the problem with the Indians especially the poor class is that they can neither grow nor have the capacity to buy food, because they are either landless or asset-less. The government's own Food Corporation of India godowns are full, and bad planning in distribution results in loss of quality of the grains in storage.

Famine is present in this century in countries like Africa. Unless proper action is taken and at the right time, it may cause widespread disease and premature death. This is presented in a recent review by Baro and Deubel (2006). According to them famine occurrence is linked to several factors like contemporary socio-economic conditions which have increased with time the vulnerability of African people to hunger, and reduced their reactions to the economic and political conflicts. To save them from famine it should first look beyond the temporary relief through emergency supplies and examine the causes of the imminent famine which remains endemic. Food security in Africa should be based on a integrated long-term solution taken up by the governments, civil society and intenational organizations which should introduce new development technologies for increasing food production and eradicating diseases.

3. PUNJAB CRISIS AND SUICIDES OF FARMERS

Sandeep Bhushan (2006) writes for NDYV about the farmers taking to suicides in many parts of

Punjab which was hitherto considered the granary of India for wheat production and it is this state which made all efforts to successfully bring about the "Green Revolution" in the 60s and 70s. The causes are discussed as mainly due to the reining of money lenders. Panjab trader-money lender is the one stop shop for meeting the farmers' needs. "Chronic indebtedness, mainly because of the vicious circle of high input costs and declining productivity of the soil, leads to land, changing hands from the farmer to the *artiya*" writes Bhushan. "In the last 15 years, 49 Jat Sikh farmers in Bhullan village of Punjab's Sangrur district have committed suicide, leaving behind their women, children and staggering loans". Sandeep Bhushan (May 17, 2006) reports that the widows of the husbands who committed suicides were asking for help to educate their children. According to his report "ever since cotton became the favored crop along with wheat in Punjab two decades ago, excessive pesticide use, growing salinity and a ballooning debt trap have led to a major shift in land ownership in the country's grain bowl". Uma Sudhir (Dec 24,2001) reported that over 300 farmers committed suicide because their cotton crop failed in spite of using pesticides. Nirpur Basu (April 28, 2002) reported several reasons for the farmers' suicide in Karnataka. Saund Singh, a 60-year-old Dalit Sikh farmer, got only 15 days of work. "There's no work here whatsoever. I will have to look for other ways to earn money," said Singh. Saund Singh's sons, Mahinder, Jagraj and Inderjeet, dreamt of working on their own and rather than those of the Jats. They borrowed money and took land on contract. However, they were unable to repay the loan and over five years, all three killed themselves.

Several hundreds of farmers have put an end to their own lives in the last few years because they saw no end to their agriculture woes.

This is the state of affairs in many states in India, though we gloat ourselves that we have solved our food problem. From the years after India attained freedom, we felt the after-effects of second world war. Agriculture, which was neglected comparatively till then, got a fillip. Fertilizers were first imported from different countries like USA and Russia. Many irrigation projects were initiated and in the next decade, more irrigated lands were brought under plough. Organic farming which is being spoken of so highly now, was actually practiced after the independence and before that. "Sheep-penning" and "cattle-penning" were very common, so also growing of leguminous grain crops like black grams and green grams in the 'rice fallows'. Similar conditions existed in England during 1500 to 1850 (Overton, 2007) as described thus, "Available nitrogen was conserved by feeding bullocks in stalls, collecting their manure (which is rich in nitrogen), and placing it where it was needed. Also, most importantly, new nitrogen was added to the soil using legumes - a class of plants that have bacteria attached to their roots, which convert atmospheric nitrogen into nitrates in the soil that can be used by whatever plants are grown there in the following few years".

Now with the introduction of mechanical farming, cattle population dwindled or even wiped out, resulting in no 'cattle, no sheep penning'. Cattle manure is not available so much now, being limited to that from the milch cows. In the last 2 decades, the poultry farming (hens grown in cages) have sprung up, instead. Mechanization of farming is no doubt necessary in terms of efficiency and bringing more land under plough, but its effects on the traditional farming with natural organic fertilization have made it necessary to depend on inorganic fertilizers which in turn lead to the soil degradation and health hazards using the grains produced by these. The "Green Revolution" no doubt was helpful in a large measure to tide over the food shortage. However, Punjab which we could as well call as "Mother" of green revolution, had all the facilities, namely, plenty of water to irrigate the crops, fertilizers supplied at subsidized price was available in enough quantities, and all that was needed was some wheat cultivars which could respond to heavy fertilization without lodging. This was provided through the wheat varieties got from Mexico and experimented in Indian farms for their performance. Now that wave of production is succeeded by situations where water is scanty, soil turned saline due to over-fertilization, and prevalence of drought. Added to these, the farmers had to heavily borrow from the money-lenders, and unable to return the money led to their suicide. In Maharashtra, there are rich landlords who are sugarcane farmers and owners of sugar factories. But suicides, drought and debts do not bother the millionaires who are actually makers of farm machines. Quite a few of them own luxury cars like Mercedes. Kolhapur is now number two, in terms of per capita income. The prosperity of Maharashtra is due to large or largest number of sugar-factories in the country. But the farmers who grow other crops like cotton have been forced to leave farming because of poor returns. The NDTV reporter Priyanka Kakodkar and Anupama

Ramachandra (January 24, 2006) wrote "The reason behind the extreme cases (of suicides) is because most of the state's resources are drained by Western Maharashtra, which is the political bastion of the Congress and the NCP.... That's why, the sugarcane fields of Western Maharashtra, that account for only four per cent of Maharashtra's farming land get 60 per cent of its irrigation water...... But the bigger scam is thousands of crores are poured into sugar mills that are controlled by politicians from the ruling Congress-NCP. .... This is why despite all the money and the water, small farmers in Western Maharashtra are in distress, just like their counterparts in Vidharbha". There is another report on the causes of suicides in Andhra Pradesh. "In the 1980s, when cotton farmer suicides were reported in Guntur district of Andhra Pradesh, synthetic pyrethroids were brought in as the solution. In 1997, when Warangal cotton farmers committed suicide, genetically modified Bt cotton was touted as a solution. Four years after Bt cotton was introduced in Andhra Pradesh, the solution seems to have become part of the problem", writes NDTV correspondent in January 24, 2006 from Warangal. Chandraiah's last hope had been the Bt cotton he grew on land that he took on rent. Everyone in his native village opted for Bt cotton this year even though the seeds were almost four times the usual cost, because they were told there would be no pests. But pests destroyed the crop and no one got beyond five quintals against the promised 10-15 quintals an acre, and that too, only after spraying pesticides. "They said there is no need to spray pesticides on Bt cotton. But these pests came. The rain also spoilt the crop. Now all is gone," said Yelliah, a farmer.

4. POPULATION AND ENVIRONMENT

The exponential growth of population and its assault on the environment have serious impact on food production. It also threatens social and political stability in the developing countries while the environmental issues are turning out as a global problem. One such environmental problem is the flooding which is result of constant deforestation. Overlogging, one cause of deforestation, is proportional to the demand from the growing population and their needs for building material, firewood, an increase in farmland and capital; the last comes from the foreign agencies, as aid to the developing countries. This was what happened in Thailand and Malaysia which depend on timber as an important source of employment and foreign exchange; these countries incurred heavy loss due to recent floods caused by deforestation.

Population and its effects on the rural and the urban areas

High birth rates in the countryside force the subsistence farmers to cultivate the marginal lands hitherto not so productive as the fertile ones. In Rajasthan, arid soils are depleted of nutrients by the intensive cultivation. The destruction of millions of acres of rain forest in Brazil in recent years by the farmers in order to eke out a living is a case in point. The increase in population in the rural areas causes exodus to the nearby and far-off cities. Consequent on the better modes of transport, the rapidity with which the cities grow is unprecedented. A six-fold increase in urban population was anticipated for the world between 1950 and 2020, and as a result the growth of cities is no longer related to the level of development. In the 1950s the urban population was 17%, and is expected to rise to 50% in 2020, in the developing countries (Keyfitz 1989).

The concentration of people in the cities and the urbanization have certain advantages, like better health care and other amenities as compared to their cousins in the rural areas. They also have no direct effect on the forests, the wildlife or the oceans, and the biosphere in general. However, there is one feature which has negative effects. People living in the cities are on the move incessantly—as commuters for vacations, business or pleasure, and their travel is by car, bus and plane which causes much damage to the ecosphere. As an example, a middle-class American eats more than their Asian counterpart, owns more clothes and has more varied entertainment. However, much of the damage to the ecosphere is related to movement and travel, and not the eating habits or the living standards.

There are 500 million registered automobiles in all the countries put together, in the 1980s and on the average they burn up nearly 2 gallons of fuel a day, and most of this occurs among the 1.2 billion people in the developed countries. It is also predicted that in the future most of the net growth in the use of motor vehicles would take place in the developing countries (Keyfitz 1989). To say it simply, the number of automobiles is increasing more quickly than the population itself, and at the current rate of increase, there will be four times as many automobiles as there are in the 1980s. All these conditions present serious threats especially for the developing countries. There is an interesting report in the News from the Worldwatch Institute on urban sprawl (Message No. 24 July 2, 2001). The United States estimated to have "the world's most car-reliant cities, and US drivers consume roughly 43% of the world's gasoline to propel less than 5% of the world's population".

The absolute increase in population coupled with such trends as urbanization, and greater mobility, with the increase in the use of automobiles, poses serious threat. To leave it to natural constraints to intervene and limit the population is to only accept famine, low living standards, unemployment, ravage by diseases and ecological disaster. Such options are not acceptable to society. Therefore the curbing of population is imperative. It is heartening that some Asian nations—China, Indonesia, Thailand and South Korea— have established family planning programs in the 1960s and the birth rates have declined from 25 to 60% over two decades. Most of the developing countries are now aware that the pace of development would be faster and the destruction of the environment slower if their populations were under control.

5. FOOD PRODUCTION AND INCREASING POPULATION

Writing about the food and population, Revelle (1974) made a comparison between the developed and the developing or underdeveloped countries. World production of cereal grains more than doubled from 1951 level in 1971, while the population increased by about 50%. However, this increase was not shared equally by the world's population. Actually more than half of it was shared by the richest 30%, while less than half reached the poorest 70% of mankind, which constitute 2.6 billion of Asia, Africa and Latin America. In real terms, the volume of food produced was barely enough for the population. The situation deteriorated in 1972 and 1973, because of large scale droughts, rise in petroleum prices and a worldwide shortage of nitrogenous fertilizers. Drastic increase in world oil price, affected the cost of pumping water for irrigation. There are no two opinions about the need for controlling population, which is the most difficult one. Food production has to be increased and it is imperative that agriculture is modernized in the developing countries through intensification of research in agricultural and food front. In the long run, the solutions for the developing countries remain two-fold as stated earlier, viz., (i) a fast reduction in population growth and (ii) a sharp increase in food production. The physical possibilities for the latter are great, in both of natural resources and of agricultural technology. When Thomas Malthus (1798) announced his "Principle of Population", his concept was limited with respect to the resources available for agriculture, viz., primarily land, water and, both human and animal labor. This is largely true for the poor nations. For developed countries however, agriculture and procurement of food remained as smaller components of their total economic activity. Household expenditure in the U.S. takes less than 13% of their general income in spite of their extravagant and rich diet. The transformation was mainly brought about by the large-scale application in agriculture of resources, which Malthus was not aware of, viz., mechanical energy from fossil fuels, and scientific and technological advances made in the last 2 decades (Revelle 1974).

The theory of Malthus has been discussed by Hansen and Prescott (2002). They have developed a unified growth theory which accounted for constant living standards prior to 1800 and subsequent ones driven by modern industrial economies. The first one is classified by them as *Malhus technology* which requires land, labor and reproducible capital as inputs. The second one is recognized as the *Solow technology* (Solow, 1957) which does not require land. In the early stages of development namely the *Malthus technology*, living standards were stagnant due to higher population growth, despite the technological progress achieved. The technological progress on the other hand, caused the *Solow technology* more profitable. Furthermore, the *Malthus technology* gives constant returns from the land, labor and capital inputs. In the *Solow technology*, only labor and capital are used. Production in this latter sector is carried out in factories, where entry or exit occurs depending on the profitability of operating an additional factory. In the *Malthus technology* the stock of useable knowledge is small, and as the population grows, the living standard remains constant. As the useable knowledge grows, the *Solow technology* begins to operate. At this point, the living standards improve so that the population growth has less influence on the per capita income. Thus the fact that the growing of living standards depends only on shifting away from the land-intensive *Malthus technology* to one that is capital intensive, and this is influenced by the rate of technological progress in the *Solow* sector. The green revolution, or any revolution is a result of advancement of scientific knowledge which were not foreseen when the Malthusian theory was proposed. However, there results a stagnation in food production until a new 'break-through' is attained. Hansen and Prescott (2002) have quoted the conclusions of Mokyr (1990) that technological advancement did not begin with the industrial revolution. But once the *Solow technology* began to be used, the advantages of "learning by doing" and more economic payoff increased the rate of technological growth.

In the 1960s, it was predicted that food production would be in shortage to feed the increasing population. However, unforeseen progress in plant science, the unprecedented use of fertilizers, and expanded use of irrigation not only solved the food problem but outpaced the population. Forty years ago, 25% of the world's population went to bed hungry each day. However, increase in food production(green revolution) reduced that percentage to about 17% even as population increased. Yet, today, 40 years later, chronically undernourished people are about 0.8 to 1.0 billion i.e., those who consume fewer than 2,000 calories per day, and there is widespread deficiency of vitamin A and iron among the preschool children and women (Conway and Toenniessen, 1999)

The relation between food production and population is simple, the larger the population the greater should be the food production. However, apparently few people anywhere and at any time have allowed themselves to live very long at the Malthusian level of "bare subsistence, i.e., food supply just sufficient to sustain life. It is an important point to note that unless there is an extreme shortage of food, most adult human beings do not die due to the inadequate amount of food. But this situation results in reduction in vitality and general health, which lead to their inability to work and play like the healthy people. Undernourished children become easily susceptible to the diseases and poor mental development. Nutrition affects the reproductive ability also. In discussing the food needs of people, the physiological requirements especially for calories, protein, vitamins and minor nutrients must be taken into account.

Revelle further states that the food demand in the poor countries is relatively insensitive to food prices, and poor families make adjustments in their budget, by forgoing other wants. If food reserves reach a low level, increasing the incentives to farmers does not solve the food problem and a slight decrease in food supply causes a sharp rise in prices. Drought and poor weather during 1972-1973 caused a drop in the world cereal production for the first time after 20 years and the cereal reserves decreased to a dangerously low level. Many planners advocated creating world food bank to be managed by international organizations.

The application of science and technology to agriculture greatly involves the knowledge of physics and chemistry of soil and water, and the genetics, physiology and plant pathology. Today the science of genetics has become highly specialized changing from simple plant breeding and hybridization to the introduction of new beneficial genes and evolving of genetically modified plants which have high yielding traits, disease resistance and so on. Looking at what we can do to fully exploit the soil resources, we can examine the total cultivated area of land on the earth which is needed to feed the growing population with the use of modern agricultural technology. With the population of the world as on 1973, which was 3.8 billion, 158 million hectares would be required under one crop per year farming. The land cultivated was 1.4 billion hectares and this is under low level of agricultural technology, in most of the world. When under high-technology farming 6 metric tons of cereal grains are obtained, the average Indian or Pakistani farmer could produce only about a ton of wheat or rice (Revelle, 1974). Because of the low photosynthetic efficiency of plants for conversion of light energy, larger areas of crops are required to produce a given amount of grains. Large areas of the earth's surface are not now cultivated for want of adequate capital and other inputs besides high agricultural technology. There are limitations set by climate, soil physical characteristics and water supply.

6. LIMITING WATER: "*WATER, WATER EVERYWHERE: NOR ANY DROP TO DRINK*" (RHYME OF THE ANCIENT MARINER)

Falkenmark (1997) has presented certain interesting facts about water needs which were subject of discussion in the transactions of the Royal society with others namely, J. W. Kijne (International Irrigation Management Institute), B. Taron (Institute of Soils and Water, Israel), M. V. K. Sivakumar (WMO), Eric Craswell (IBSRAM), Bangkok. Incidentally Malin Falkenmark was the first recipient of Henry Darcy medal, 1999, "in recognition of her pioneering work in water resources assessment and her innovative approaches to the sustainable use of water resources".

The basic issue related to the ability of soil and water resources to provide enough water for the future population in the year 2025. It deliberated on the fresh water in particular, and its interaction in crop production between soil moisture and water in aquifers which are referred to as 'green water' and 'blue water' respectively. Water availability in the root zone is important for photosynthesis which consumes enormous quantities of water including that lost through transpiration. Furthermore, it is the 'blue water' which contributes largely to the increased crop production for the future. The likely increase in world population, is estimated to be 86 millions each year which is equivalent to feeding an additional India for every decade. The author states historically there were migrations of people from the dry climate regions to that having good water, as is seen in the ancient civilizations of the river basins. Another solution will be to ship food, grown in water sufficient countries, to the low income, food-deficient countries in the dry regions of Africa and Asia. The title of the discussion was "Are we on the Malthusian precipice?" The discussion ended with certain suggestions. The food-deficient countries will have to change the food self-sufficiency goal. And, the world's "food basket" will lie in sub-humid and humid regions or in other words, the affluent and well-developed nations. These regions have to change their agricultural policies and intensify crop production (Falkenmark, 1997).

Water must be available during the growing season in amounts sufficient to compensate the evaporation from the soil and the plant transpiration. Revelle quotes the 1967 report of the President Johnson's Science Advisory Committee, *The World Food Problem*, according to which the world area of potentially arable land is 3.2 billion hectares, which is 24% of the land area of the earth. 500 million *ha* of this are in the humid tropics, where rainfall exceeds evaporation and no technology is available to intensify and use this precipitation. Another 300 million *ha* could be harnessed if water is available for irrigation. On an examination of the flow of water from the world's rivers, that could be used for irrigation, only a fraction of the runoff is utilized for the farms. In other words, less than 4% of the total river flow is utilized for irrigating 160 million *ha*. Thus the potential for exploiting the river waters is very high, the only limitation being the uneven distribution of river water flowing across different countries, or parts of the same country as in India, Pakistan and Bangladesh. This results in the utilization of about 30% of the land that could be brought under irrigated cultivation. Any major extension of the cultivated area at present, needs a huge capital investment. The uneven distribution of people is a greater obstacle than the uneven distribution of the potentially arable land. Seventy percent of the world's population are in Asia and Europe, where nearly all the potentially arable land is already under cultivation. Large scale irrigation schemes are required to bring under plow the potentially arable land. If the people in Asia need to have sufficient food in the future, it is necessary to increase the yields on the presently cultivated land, by growing two or three crops year, and this would call for extensive irrigation development according to Revelle.

When we consider the building of large dams across big rivers, there are certain factors to be fully examined. Although such dams would be very useful in diverting the water for irrigation agriculture from being lost into the sea, the hazards on the environment and the displacement of people from the areas should be prevented. There has arisen big controversy about the building of the dam across Narmada river and the world commission of dams has advised the re-evaluation of the benefits and if these large dams could really provide water as well as energy through the hydel projects. WCD reports that contrary to accepted belief, the reservoirs of large dams release green-house gases which are as damaging the environment as thermal power stations.

7. WATER HARVESTING——MEANS TO CONSERVE WATER

The availability of adequate and good quality water is essential to the existence and prosperity of a nation. In many regions of the world, rain water is scanty and therefore attempts to store it in different storage systems like tanks and large wells have been successful hitherto. But with the increasing population, water requirement has increased, while the rainfall is getting less, especially in the tropical areas. One method to conserve water is to follow rain-harvesting techniques, and introduce strict measures to careful use of water. It is reported that nearly 1.4 billion people live in regions which will have severe water shortages in the first 25 years of this millennium. These regions do not have sufficient water resources for irrigation, and also do not have enough water for use in domestic purposes, in particular. Industrial requirement of water is large in these developing countries which are engaged in starting new or expanding the old industries which help produce consumables for local and export market. The IWMI (International Water Management Institute) studied the water supply and demand of 118 countries for the periods from 1990 to 2025, and projected a situation of declining water tables in the semi-arid regions of Asia.

The developing countries are engaged in a desperate race to keep food supplies enough to feed the population. During 1951 to 1971, world production of cereal grains nearly doubled while the population increased by 50%. There was substantial rise in cereal consumption by nearly 40%, although more than half of it was consumed by the richest 30% and the 70% of population shared the remaining less than the fifty percent. The situation is said to have deteriorated in 1972 and 1973, due to droughts in many countries, including India, and Africa. The limiting factors in crop production and agricultural development in Africa and South America in general, are not natural resources, but economic, institutional and sociopolitical constraints. Revelle's projection shows that the quantity of potentially arable land on the earth is considerably larger than available now for cultivation. In the developing and less developed countries, agriculture is an important component and modernization of agriculture is essential for overall economic development. This in turn depends on many inputs from outside agriculture as well as a large and growing market for the products (Revelle, 1974).

Ecological deterioration will be on the increase following the extension of agricultural activities from the fertile land to the marginal and unsuitable soils. Widespread use of pesticides and over-fertilization would increase the environmental damage. Some people have called for a return to the old methods like bullock drawn plows and grinding human labor as the main form of energy used in the farming. However, if the growing population of Asia are to be fed, this reversion is not possible in the realities of energy requirement and extensive farming. One can have a reasonable idea if the energy utilized in U.S. and India as an example. Revelle has quoted the work of others who have estimated that the solar energy captured in the grains of corn is two and a half times the total energy from fossil-fuel used in US corn production including that used in manufacturing farm machinery, chemical fertilizers and their transportation etc. However, this low energy input is due to the fact only 4% of the corn need irrigation. In comparison, the ratio of total energy required to food energy produced is 0.525 for a modern irrigated agriculture and food-processing system in India. There is therefore a need to accelerate the rate of modernization of Indian agriculture. The Indian Irrigation Commission has projected a doubling up of the irrigated area from 43 million cropped hectares in 1973-74, in the next 30 years, which is expected to cost $ 14 billion, and is only 1% of India's gross national product. If the projected irrigation development is combined with an optimum utilization of fertilizers, also using highly fertilizer responsive crop cultivars having high yield potentials, the problem of food supply in India can be pushed to the background for several years to come.

Leaving aside for a moment, the picture relating to the drinking water in cities like Delhi is appalling. Newspapers are full of stories how the river waters are polluted by discharge of industrial wastes and town sewerage. The most revered Yamuna river which meanders south to cities like Mathura, and Agra, it is the principal source of drinking water, too. The well-waters in many cities have dried up or become salty. The reports tell that the city's water crisis is not a question of lacking in water, but the problem of distribution which is hampered by a feeble infrastructure and a lack of resources.

India has many rivers and some of those in the Northern India are perennial. The Ganges begins from the Himalayan region, and runs through several states to finally confluence in the Bay of Bengal. The river Yamuna is or at least was a perennial one. But today it does not carry either potable or irrigable water, since it is all polluted by the discharge of wastes from the industries set on the banks. It may surprise one who could know that it was the river on the banks of which the mythological Krishna and Gopikas were dancing and playing in the large gardens or brindavans. Now none would think that such thing ever happened, if real. Another river Saraswathi has also become mythological, although there is some evidence based on satellite scanning, that this was existing once upon a time, and has disappeared, perhaps due to human exploitation. Robert Glennon (2002) has described how the Santa Cruz river went dry and also about the fate of other rivers like The Upper San Pedro River in Arizona. Gold mining industry in Nevada have caused water shortage due to excessive pumping of water deep below the ground while mining for gold ore. The use of bottled drinkable water started originally in USA and Eric Carlson teamed up with Hugh Hastings to supply spring water to a water company under the label of Ice Mountain (Glennon, 2002).

In the editorial of California Agriculture (2007), the quality of surface waters is discussed. It is regulated by the US Environmental Protection Agency and the State Water Resources control board. The federal government is responsible to regulate the 'point-source' pollution while the local and state governments take control of nonpoint-source pollution through regional plants. They also use indicator bacteria such as *Escherichia coli*, for maintaining the purity of water. *E. coli* is harmful to health if it exceeds a certain limit. Water quality often intersects with human health concerns. The spinach crop gets contaminated with *E. coli.* and this spinach outbreak forced the government to establish the Center for Produce Safety. The Center, will help prevent the health risks caused by a virulent strain of *E. coli.* A body is formed with the signing of Leafy Greens Marketing Agreement which certifies adherence to practices that reduce the risk of waterborne diseases. Growers are advised to avoid letting water stand in fields because it can attract animals, avoid the use of storm water for irrigation because it can have high levels of bacteria. Such stringent measures are required for heavily populated countries like India, where the river waters are indiscriminately used and polluted in so many ways.

Furthermore, there is a problem with regard to the perennial water source in California. Factors which affect are drought in many parts of the state, increasing demands from the population and the deteriorating conditions of the San Francisco Bay-Delta infrastructure. The federal court has ordered to dramatically reduce water deliveries to agriculture and residences. The UC researchers are devising measures to control runoff pollution from farms, nurseries and cities, since the runoff could carry pathogens, pesticides, nutrients and sediments (editorial California Agri 2007).

Another instance where a big canal was made in the name of Prime Minister Indira Gandhi is the water being neither useful to the farmers nor for the people's drinking purpose. *Ghai (2005) reports that despite the development of the canal system for Rajasthan, the lives of communities in this region of Thar, suffer from severe droughts year after year. In the year 2005, the government declared that 75% of the IGNP in Bikaner and Jaisalmer as affected by drought. He has analyzed the various causes for the ineffectiveness to save the area from drought. One main reason is the "peculiar credit and speculation-oriented cash economy" which have increased the debt burden of the farmers in purchasing agricultural equipments, fertilizers etc. Despite this kind of borrowing, the farmers did not get enough water for the crops. The cost of maintaining the land for crop production has soared high and farmers find it uneconomical to do farming. Because of the drought the labor force prefers to migrate to neighboring districts. Though the IGNP was intended to encourage growing of cash crops which need good irrigation, the farmers could not realize this, due to poor water supply in the IGNP.* "Still, the IGNP continues to be presented as a perennial canal system. The present water crisis has been explained by the 'failure of rains' in Himachal and Punjab or by giving pleas about discords in inter-state water sharing. No doubt, there is more than a grain of truth in these meteorological observations, but ascribing these for the water crisis looks a bit too preposterous", says the author.

Global freshwater availability has not been assessed so far. Rockstrom *et al.* (1999) have made a detailed study on several aspects relating to the atmospheric water vapor, and water requirement or its utilization for food production and the ecological systems. They have claimed it as the first "bottom-up estimate of continental water vapor flows". Accordingly the freshwater requirement for the increasing demands for food production at the projected year 2025, there is a 'trade-off' between the crop requirement and that needed for the maintenance of the life-sustaining ecosystems. The authors have cited the work of several others who have stated that freshwater availability for the needs, like the industries, household, and the irrigation of agricultural crops, pose serious limitation. About 25% of the world's crops face water scarcity thus creating the need for import of food. Water quality also would deteriorate thus reducing the availability of good water for the people. There are three options for the water conservation. The first proposal according to the world organizations like FAO, UNDP, advises to increase the irrigated agriculture to combat food shortage. Shiklomanov (1997) has estimated that the evapotranspiration caused by irrigated agriculture by 2025 would rise to 425Km^3 per year or about 14% additional freshwater requirement. The second proposal is to increase rain-fed agriculture, that is, growing crops with rain-water (Falkenmark *et al.* 1998). There are two pre-requisites for this. One is, to improve the WUE (water use efficiency) of crop species, and in the words of the authors, "*redirecting in-field evaporation to transpiration* within croplands, i.e., increasing the yields with the same amount of water vapor flow". It implies selection of crop varieties which are efficient in water use, improved farming practices and soil management to conserve water. The other avenue is to "*redirect evaporating surface runoff for use in croplands*". The third proposal is in the words of the authors, "*redirect substantial amounts of water vapor flows from other biomes to cropland*".

With reference to the second proposal by Falkenmark, that the crops have to be grown under rain-fed conditions, Fereres and Soriano (2007) are of the view that in the coming decades, irrigated agriculture will have to be under conditions of water scarcity which means it has to be brought down. There will be insufficient water available for growing crops under irrigation and crop production will be reckoned on the basis of unit of grain yield per unit of water consumed. To cope with scarce supplies, deficit irrigation, has to be followed which means application of water below full crop-water requirement. While "deficit irrigation" is in vogue over millions of hectares for a number of reasons—it has not received sufficient attention in research. Several cases on the successful use of regulated deficit irrigation (RDI) in fruit trees and vines have been recorded, showing that RDI not only increases water productivity, but also farmers' revenue. Research linking the physiological basis of these responses to the design of RDI strategies needs to be intensified and adopted in water-limited areas. As stated earlier by Falkenmark *et al.* (1998) evolving of crop cultivars which are water use efficient is important. In the IRRI at Philippines, Lafitte *et al.* (2007) have examined some varieties of rice. The high-yielders of rice are developed for growing under irrigation and do not withstand drought as in the case of the lowland cultivar IR-64 grown widely. They are considering superior high yielding lines to be backcrossed with ones, which have drought tolerance and then make selections. They also use a criterion for drought resistance in terms of their response to ABA and ethylene treatments.

Falkenmark (1997) has discussed the water requirements for the growing population of the world during the years 2025. Water supplied at the root zone, and he calls this as 'green water', is important for crop production and this is greatly deficient in the developing countries. Under deficient conditions, the run off water which he calls as 'blue water' could be provided to the crop. The water deficit is experienced in drier regions in Asia and Africa in particular. Water requirement in these regions will increase over the years and about 75%

of the world population will lie in these countries by the year 2025. The increase in water demand would be as a result of increased population as well as that needed for more food production through irrigation. Raising the question, "Are we on the Malthusian precipice?" a few solutions have been proposed. The vulnerable countries will have to give up the food self sufficiency goal for a few years, since the problem is so critical that cannot wait for policy decisions. Alternate measure to combat drought is to develop technologies to reduce the irrigated farming, and concentrate rain-fed agriculture. In order to avert the 'Malthusian precipice', there is need for an urgent change in the global agricultural policy and trade.

With increasing demand for water for irrigation, there is great concern about its careful use for both irrigated and rainfed crops. WUE (water use efficiency) is the amount of biomass or grain yield produced by a unit of water consumed. Once this is derived, farming system could be modified through efficient soil management practice. This includes the control of loss of moisture through surface evaporation and plant transpiration. Hatfield *et al.* (2001) have discussed the role of several factors. Water use is influenced by the nutrients like N and P. It is possible to increase the WUE upto 40% through soil management. An increase of upto 25% is possible through efficient nutrient management. Irrigation of crops is required for high yields and is a vital component of agriculture. It is estimated that it is followed in about 263 million hectares as known in 1996 and 49% of this area lies in India, China and USA. Howell (2001) has discussed some important aspects of irrigation in relation to the needs for food production and various concepts of WUE (water use efficiency) and its impact on water conservation.

When water shortage occurs as it happens with poor rainfall or increased use of ground water, as also the evaporation of water due to high atmospheric temperature, the drought occurs. Crop yield is reduced. Passioura (2007) has discussed the different kinds of drought and how it affects crop production as well as how the farmers need to combat to produce crops with good yields. While agronomists define drought as that which is caused by too little water, to a molecular biologist, it is sudden and severe water deficit. There are a few situations which are caused by drought. The soil water is lost rapidly through evaporation and plant transpiration. Water deficits as defined by the farmers and agronomists create different problems, affecting plants at the various development stages, from the time of sowing, to flowering and grain filling. How frequent the crops are affected depends on the vagaries of the climate or actual rainfall. Farmers do have certain techniques to combat drought. They resort to sowing the seeds deep in the soil so that if the surface is dry, the seeds could use the moisture below. However, it depends on the types of seeds, as for example, the semi-dwarf wheat would not be able to grow and reach the surface above the soil, if planted deep. Rebetzke *et al.* (2007) have proposed measures to solve this, by evolving cultivars of wheat which are semi-dwarf, but have long coleoptiles to come through the soil. Passioura thus emphasizes that the crop physiologists and plant breeders could work together to evolve crop species to fit in the environment and water limitations. He cites the work of Richards (1991) "In Mediterranean environments for example, getting the time of flowering of crops right is especially critical—early enough to avoid the worst of the heat and large evaporative demands of late spring and summer, yet late enough to avoid major risk of frost damage at flowering". In this effort to combat fluctuating weather conditions, the scientists perhaps should work along with the climatologist or meteorologist in the region.

8. NEW IRRIGATION TECHNOLOGY

The development of drip-irrigation technology in 1960s opened the doors for the greatest advancement in fertilizer management especially for vegetables. Early work for example, in Israel (Bar-Yosef, 1977) and later in other countries formed some of the basis for the water-management technology that is used today. Drip irrigation led to improvements in water application efficiencies and also reduced the amount of water used for many vegetables by 50% to 70% (Clark, 1992). Furthermore, the reduction in water application has positive implications for nutrient efficiencies, especially N since it is closely related to irrigation efficiency in sandy soil production areas (Hochmuth, 2000). Fertilizer materials could be injected into the drip irrigation system, a process referred to as 'fertigation'. Soluble nutrients like N and K are supplied through drip irrigation by injecting small amounts and is also regulated according to the seasonal crop requirement. Schedules for such injections were developed in Florida and other southern states (Cook and Sanders, 1991). Water conservation technology has also been developed using many advanced technologies and remote monitoring with computers. One recent advancement is described below.

Writing on the ARS role to help conserve the vital resource for agriculture, Robert Evans (2007) has presented a new system called 'Wireless Watering'. He states that worldwide, irrigation is the largest single consumer of fresh water, using up to 60% of this precious resource, and most of this are for growing food, animal feed, fiber and fuel crops. At the annual meeting of the AAAS held in San Francisco in February 2007, Evans and E. John Sadler both agricultural engineers discussed their latest findings. He has used the latest wireless technology to boost irrigation efficiency. With a wireless network of small soil-moisture and temperature sensors, the exact water needs of the crop are continuously monitored and the signals sent to the irrigation station direct the individual sprinklers to spray only that much optimum water and to the exact area for supplying the water. Evans and James Kim who designed and integrated the wireless components, are still evaluating the full benefit of this systems on the water saving, and also conserving fertilizers. This will help reduce the pollution of the ground water and soil. They further state that if irrigation farming is to be profitable in future, such kind of innovations is important. Evans says, "Before, we were focused on how much productivity we could extract from a unit of land, now it is time to start thinking about how much we can wring from each unit of water consumed during the production process".

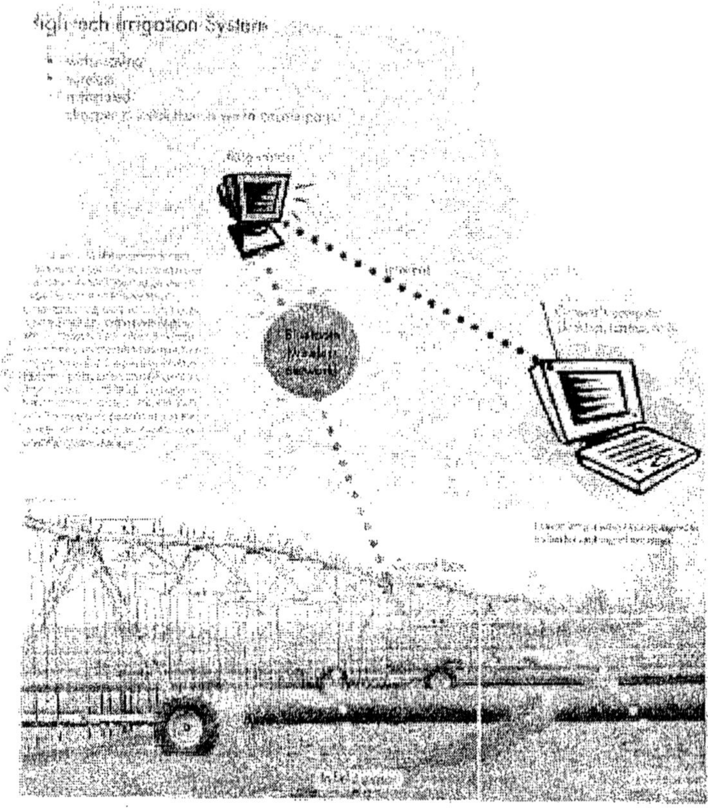

PICTURE 1 (WIRELESS WATERING)

In-field sensing stations monitor soil moisture and soil and air temperatures. A weather station mounted on the irrigation cart monitors weather conditions in the field. All in-field data is sent wirelessly to the base station which processes the information and determines whether and how much to irrigate based on the precise location (determined by differential (GPS) of the irrigation cart in the field. The base station communicates back and forth with the mobile irrigation cart and the grower, who has ultimate control over what the station does.

(Credit: Robert Evans, USDA-ARS Northern Plains Agricultural Research Laboratory, 1500 North Central Ave.,m Sidney, MT 433-5038. Picture from Agricultural Resarch July 2007 page 6).

Crops like rice are essentially grown under semi-aquatic condition and the seedlings are transplanted in the puddle field. Large amount of water and labor are thus required for growing wetland paddy crop. The alternative methods for rice growing have been examined by Bhushan *et al.* (2007). A field experiment was conducted in the Indo gangetic plains for 2 yr to evaluate various tillage methods for their efficiency in, water, labor and energy use. Rice was grown in the puddle, and direct-seeding systems. The yields of rice in the conventional puddled transplanting and direct-seeding on puddled or nonpuddled (no-tillage) flat bed systems were found to be equal. Normally, puddled transplanting required 35 to 40% more irrigation water than no-tillage direct-seeded rice. Compared with conventional puddled transplanting, direct seeding of rice on raised beds had a 13 to 23% savings of irrigation water, but with an associated yield loss of 14 to 25%. Nevertheless, water use efficiency (WUE) in the rice–wheat system was higher (0.45 g L^{-1}) with direct-seeded rice than with transplanted rice (0.37–0.43 g L^{-1}). In Year 1, no-tillage rice–wheat had a higher net return than the conventional system. But in Year 2, the net returns were equal. The results led to the conclusion that the conventional practice of puddled transplanting is better than no-tillage-based crop establishment methods which in addition to labor saving also help save water. The effects depend on the amount and distribution of rainfall during the cropping season, since the water saving is beneficial only under no rainfall or poor rainfall.

Water situation all over the world is increasingly becoming grave. It is interesting to know that there will likely be a war on the use of water at some future date. John Flesher writes in *Chicago Sun Times September 11, 2006* that the Aral Sea is now reduced to a quarter of its previous area after 50 years, and this is the result of diversion of the rivers which were feeding the sea, for agriculture in the central arid regions, by the then Soviet regime. Peter Annin, a former *Newsweek* correspondent, who visited the region for collecting information for his book, *The Great Lakes Water Wars*, under publication by Island Press, remarked that 1/5th of fresh surface water "It kind of defies the bounds of the mind to grasp how dire the ecological situation is there". He does not expect the ecological and political differences would make the Great Lakes suffer the Aral's sea's fate, but the tragedy of Aral sounds a note of warning for the people. Annin says: "What it showed to me in a very surreal way was that these giant lakes are vulnerable, they actually can be drained. They are not immune to human destruction." He thinks that an era of warring over the Great Lakes is under way — and will intensify when the global water shortage turns worse. "The lakes' future hangs in the balance as leaders grapple with preserving nearly one-fifth of the world's fresh surface water". The former Soviet president Mickhail Gorbachev while delivering the address at the Earth Dialogues forum in Brisbane on July 22, 2006 warned that there is danger of war due to a worldwide shortage of water and energy. The major river basins in several countries would turn the regions of military conflict, as the water for drinking and irrigation of agricultural lands would run short. Mr. Gorbachev quotes Kofi Annan's statement secretary general of the UN (United Nations), "In 2001, in the new century we may see wars over water, rather than over oil" (Roberta Mancuso AAP News, July 22, 2006).

9. GLOBAL WARMING AND CROP PRODUCTION

There have been many views and discussions on global warming. The climate alarmists were making repeated claims that there has been rise in temperature, and there will be catastrophes in future, like hurricanes, ice melting in the poles and the climate change will affect agricultural production. This is supposed to be caused by burning fossil fuels, and gasoline, coal and natural gas used for heating, cooling in winter and summer, and electrifying homes. Lindzen (2006) Alfred P. Sloan Professor of Atmospheric Science at MIT have discussed the global warming in the opinionjournal. He reports that the science of climate is very much misunderstood and the scientific statements made about climate change has been created by vested interest, thus raising stakes for policy makers to provide more funds for research in this area. In contrast, the funds for dissenting scientists are cut down, and themselves "libeled as industry stooges, scientific hacks or worse". A one-degree rise in the global mean temperature since the late 19th century is accepted as source of weather catastrophe. The fact is not understood and deliberately avoided by the press and public. The widespread scientific support is that global temperature has risen about a degree since late 19th century. The levels of CO_2 in the atmosphere have increased by about a third in the same period contributing to future warming. These are true. But these facts neither lend support for the alarm nor establish man's responsibility for the small amount of warming that has occurred. And, the loud alarm-makers are actually demonstrating skepticism of the very science they say supports them. It is not just that the alarmists are trumpeting model results which are blatantly wrong. They are prophesying catastrophes that couldn't happen even if the models were right and their justifying costly policies supposed to prevent this phenomenon. If the models are correct, global warming would bring down the temperature differences between the poles and the equator. And this difference in temperature, would actually reduce the excitation of extra tropical storms, and not increase. In this connection, it is important to note the comments made in his personal communication, Bernard Switalski wrote on December 5, 2006 the following:

(quote) (Switalski) remarked, "Don't believe what you read, Warmer cultists lie all the time. They lie, they dissemble, they dissimulate, they twist facts, and they withhold facts. And the media swear to it." Find attached a short piece I wrote a couple of years ago on the rise of Lysenkoism in the Soviet Union, and the Soviet media's assistance in creating the Lysenko monster. Re global warming, what we are witnessing today is an almost exact replication of what happened re Lysenkoism in the Soviet Union back in the 20s, 30s, 40s, and 50s. Keep all this in mind the next time you read about some zany professor who says the frogs are vanishing because of global warming. And it is because of data like these that by 1992, 4,000 scientists, including 72 Nobel laureates, had signed the Heidelberg Appeal", and another 14,000 scientists - about a third holding doctorates - have signed Frederick Seitz', "Anti-global-warming Petition". Meanwhile, in contrast to many and switalski's 'anti-global warming' group, Al Gore is awarded Nobel Prize. The following is Gore's comments:

"It is the most dangerous challenge we've ever faced, but it is also the greatest opportunity we have had to make changes," he later said at a brief news conference in Palo Alto, Calif. Gore did not take any questions. As he walked away a reporter asked if he would run for president, but Gore did not respond.

Gore's film "An Inconvenient Truth," a documentary on global warming, won an Academy Award this year. He had been widely expected to win the peace prize.

"His strong commitment, reflected in political activity, lectures, films and books, has strengthened the struggle against climate change," the Nobel citation said. "He is probably the single individual who has done most to create greater worldwide understanding of the measures that need to be adopted."

It cited Gore's awareness at an early stage "of the climatic challenges the world is facing."

REFERENCES

Baro, M. and T.F. Deubel. 2006. Persistent Hunger: Perspectives on vulnerability, Famine, and Food

Security in Sub-Saharan Africa. Ann Rev Anthropology 35:521-538.

Bar-Yosef, B. 1977. Trickel irrigation and fertilization of tomatoes in sand dunes. Water, N and distribution in the soil and uptake by plants. Agron J 69:486-491.

Bhushan, L., K., J.K. Ladha, R. K. Gupta, S. Singh, A. Tirol-Padre, Y.S. Saharawat, M. Gathala and H. Pathak. 2007. Saving of Water and Labor in a Rice–Wheat System with No-Tillage and Direct Seeding Technologies. Agron J 99:1288-1296.

Clark, G.A. 1992. Drip irrigation management and scheduling for vegetable production. HortTechnology 2:32-37.

Conway, G. and G. Toenniessen. 1999. Feeding the world in the twenty-first century. Nature Suppl 402: C55-C58.

Cook, W.P. and D.C. Sanders. 1991. Nitrogen application frequency for drip irrigated tomatoes. Hort Science 26:250-252.

Dando, W.A., Biblical Famines. 1983. 1850 B.C.-A.D. 46: Insights for Modern Mankind, Ecology of Food and Nutrition, 13:231-249.

Evans, R. 2007. Wireless watering: New irrigation technologies from ARS can help conserve a vital resource. Agricultural Research July, pp 6-7.

Falkenmark, M. 1997. Meeting water requirements of an expanding world population. Phil. Trans. R. Soc. Lond. B (1997) 352: 929-936.

Falkenmark, M., W. Klohn, J. Lundqvist, S. Postel, J. Rockström, D. Seckler, S. Hillel, and J. Wallace. 1998. Water scarcity as a key factor behind global food insecurity: Round table discussion. *Ambio* 21(2): 148 - 154.

Fereres, E . and M. A. Soriano. 2007. Deficit irrigation for reducing agricultural water use. J Expt Bot 58:147-159

Ghai, Rahul. 2005. Disaster Awaits! A sham in the name of Indira Gandhi. Tehelka. The people's paper June 18.

Glennon, Robert. 2002. Water Follies: Groundwater Pumping and the Fate of America's Fresh Waters. Island Press Publishers, Washington, Convelo and London. 314 pages.

Hansen, G.D. and E.C. Prescott. 2002. Malthus to Solow. Amer Econ Rev 92:1205-1217.

Hatfield, J.L., T.J. Sauer and J.H. Prueger. 2001. Managing soils to achieve greater water use efficiency. Agron J 93:271-280.

Hochmuth, G.J. 2000. Management of nutrients in vegetable production systems in Florida. Soil Crop Sci Fla Proc 59:11-13.

Howell, T.A. 2001. Enhancing water use efficiency in irrigated agriculture. Agron J 93:281-289.

Keyfitz, Nathan, The Growing Human Population, Sci. American 1989, sept, pp 119-126.

Lafitte, H.R., G. Yongsheng, Shi Yan and Z-K Li. 2007. Whole plant responses, key processes, and adaptation to drought stress: the case of rice. J Exp Bot 2007 58:169-175

Lindzen, R. 2006. (Alfred P. Sloan Professor of Atmospheric Science at MIT). "Global-warming alarmists intimidate dissenting scientists into silence" OpinionJournal – (Extra Wallstreet journal) April 12.

Malthus, Thomas R. 1798. An essay on the Principle of Population. Oxford: Oxford University Press. (*see* Hansen and Prescott, 2002).

Mokyr, J. 1990. The Lever of Riches: Technological Creativity and Economic Progress. Oxford and New York; Oxford University Press (*see* Hansen and Prescott, 2002).

Overton, M. 2007. Agricultural Revolution in England 1500 - 1850. BBC Home Page History, 20 May 2007.

Passioura, J. 2007. The drought environment: physical, biological and agricultural perspectives. J Exp Bot 58:113-117.

Rebetzke GJ, Richards RA, Fettell NA, Long M, Condon AG, Forrester RI, Botwright TL. 2007. Genotypic increases in coleoptile length improves stand establishment, vigour and grain yield of deep-sown wheat. Field Crops Research 100, 10–23.

Revelle, Roger, Food and Population, Scientific American, 1974, 231:161-170

Rockström, J., L. Gordon, C. Folke, M. Falkenmark, and M. Engwall. 1999. Linkages among water vapor flows, food production, and terrestrial ecosystem services. Conservation Ecology 3(2): 5. [online] URL: http://www.consecol.org/vol3/iss2/art5.

Shiklomanov, I. A.. 1997. Assessment of water resources and water availability of the world. *Background Report No.2 of the Comprehensive assessment of the freshwater resources of the world.* WMO-SEI, Stock holm, Sweden.

CHAPTER-III

FERTILIZERS

1. FOOD PRODUCTION: ITS IMPACTS ON SOIL

It was estimated that the earth's population will reach 10 billion by the year 2100 and the question is whether our species will be able to feed itself at this stage. The short answer is 'yes' according to Crosson and Rosenberg (1989). The world food production will continue to grow rather slowly, and there would still be enough food for the population. The long answer is that it is not that simple. Food supply must expand, but at the same it should not be at the destruction of the environment, and towards this end, a steady stream of new technologies that minimize contingencies like soil erosion, desertification, salinization of the soil etc. must be evolved. In many parts of the world the supply of agricultural land is threatened by various kinds of degradation. This includes erosion of soil by wind and water. Degradation of rangelands in the arid, semiarid and sub-humid regions, water-logging and salinization of irrigated lands all come under desertification. United Nations Environment Program had estimated that about 60% of the 3.3 billion hectares of agricultural land that are not in humid regions are affected to some degree of desertification in general. The U.S. has an estimate of soil erosion and its effect on crop productivity, as in 1980s. According to this, if the current rates of cropland erosion continued for 100 years, crop production will decrease by 3 to 10%. However, the yield increase achieved through technology adoption, would more than compensate for this loss. In a recent report Vance (2001) has discussed about the future food needs and the science to tackle the situation. It is stated that the world population nearly doubled to 6 billion, in 40 years since 1960 and would go to 9 billion in the next 40 years and the needed food production to be 5.5×10^9 metric tons at 2030-2040 and the cultivated land area from the present 1.5×10^9 hectares to 1.8×10^9 hectares in the next 40 years. Nitrogen fertilizers used in 1960s were 10 million metric tons, 88 in 2000, and expected to be 120 million metric tons and similarly phosphatic fertilizers as 9, 40 and 55-60 respectively for these periods. Stewart *et al.* (2005) presented data to show that crops which were not given N fertilizers produced very much less yield. They used the yield data from USDA report as the baseline yields and calculated the percentage of yield reduction. Thus, the reduction of yield when N was not given was 27% for rice and 19% for barley. The reduction was very high for corn and cotton, as 41% and 37% respectively. Their findings clearly demonstrate the importance of fertilizers in general and N in particular.

Although it was predicted in 1960s that food production would not keep pace with the population in the years 2000, with the ushering in the green revolution, food production increased more than required for the increased population. This was largely due to the extensive use of fertilizers and the greater development of irrigation system which were under progress in the 60s. This increase in the use of N and P fertilizers caused degradation of soil, air and water. Increased use of water for irrigation resulted in recycling of the aquifer and soil and water in turn became salty and brackish, affecting the crops. Now the additional usage of N and P fertilizers would have impact on sustainability and environmental quality. Vance has posed the question, "Can we feed the projected 9 billion people in 2040? Probably yes, but at an accelerated impact on sustainability and environmental quality". He recommends intensive research for using the genes for N fixation in crop plants through traditional plant breeding and transgenic technologies, and increase in the use of leguminous plants which would fix atmospheric nitrogen and enrich the soil. There should be lesser dependence of fossil fuel based fertilizers for N and P.

Legumes are plants which form nodules in the roots and fix nitrogen symbiotically. These are next in importance as human food, and provide proteins. Beans, soybeans and lentils are widely grown for the seeds which are rich in protein. In addition, these cover a number of species, which are grown for fodder and for the improvement of soil. Legumes account for 27% of the world's primary crop production and 33% of the dietary proteins are provided by grain legumes alone (Graham and Vance, 2003). More than 35% of the vegetable oils are derived from soybean and peanut.

Legume crops could be grown in soils with low N, because of their ability to fix atmospheric N. Application of fertilizer N has increased several folds in the last 50 years and this has been responsible for the success of green revolution. But fertilizer N has limitations and cause degradation of soil and environment. Excessive N application results in leaching and contamination of ground water. In this connection, it is very beneficial to introduce N to the soil by growing legume crops. The legumes also help sequestering carbon from the atmosphere through photosynthesis, and thus both N and C will be added through the legume crops. The authors strongly recommend more research on the legumes. Growing legumes for supplying N will be particularly useful and economical for small landholders, who find it difficult to buy costly inorganic fertilizers.

Rice is the staple food for over 2/3 of world population and for people in the Asian and south Asian regions. Lot of research has been carried out to increase the yield of this crop and there is a International research institute wholly devoted for rice improvement and its production was increased dramatically combined with similar increase in wheat production leading to the Green Revolution. The acute food shortage in the 60s was tided over due to this. However, food production is not commensurate with the subsequent population explosion and with the projected growth in the year 2025, plans are under way to nearly double this present output. The classical methods will however not provide the expected result. New technologies like the genetic engineering have to be exploited. For an increase in yield, nitrogen input has to be doubled. But in this approach the resulting eutrophication needs to be avoided. It is necessary to understand the processes involved in the mechanisms of N uptake for maximum yield. Britto and Kronzucker (2004) have discussed the role of bioengineering in rice for N uptake in four areas. Rice is not a legume which can fix atmospheric nitrogen through symbiosis with a number of bacterial groups known as 'diazotrophs'. The transfer of the mechanism to a 'non-symbiotic' rice plant is not feasible in the near future. Diazotrophs found in the rice fields either as free living or in association with a legume crop or the aquatic fern *azolla* carry out N fixation and can indirectly provide N for the rice crop grown in the wetlands. This has been practiced over the last few decades successfully, although the amount provided is insufficient. Other areas like enhancement of nitrogen utilization within the plant, in other words, manipulation of nitrogen metabolism to improve the efficiency and prevent the loss of N, the management practices to conserve soil nitrogen would produce desired yield increases within a reasonable time.

The challenge to agriculture is not only to provide food for the billions, but also to achieve it with less environmental damage. This twofold goal requires continued long-term support from national agricultural establishments and also that from the Consultative Group on International Agricultural Research (CGIAR) at the World Bank, Washington D.C. The CGIAR system is a major resource for research on new agricultural technologies for developing countries. The Green Revolution of the 1960s was largely the result of varieties of rice and wheat developed in the Philippines and Mexico by the research workers of the CGIAR institutes. The food production increased at the rate of 2.4% annually from the 1960s to 1980s. Meeting the goal of food supply to the growing billions substantially, a steady stream of new technologies has to come from the national institutions of the developed and developing countries. Of the technologies needed, three categories are of particular importance: those that reduce the environmental burden of pesticides and fertilizers, those that reduce the demand for irrigation water and those that continue to improve crop production per hectare. One of the main components of fertilizer is nitrogen. If plants other than legumes could be biologically engineered to 'fix' nitrogen in the soil, the demand for fertilizer will get reduced. The most suitable candidate will be corn, because of its extensive cultivation for grain production, and also because it needs large amounts of nitrogen. This can be achieved by possible manipulation of the genetic components. Although it is by no means out of reach, it is a very complex engineering job.

2. NITROGEN FERTILIZATION

The agricultural food production was doubled in the last 4 decades and this was achieved using larger amounts of N fertilizers. The crop production was nearly doubled in the last 4 decades and fertilizer use especially the nitrogenous one increased 7-folds (Hirel *et al.*, 2007). However, these have been used excessively in our efforts to feed the growing population, not being aware of the resulting eutrophication. With the higher yields obtained, the fertilizers also caused degradation of soil and water quality. The excessive N applied leaches through and increases the N content of aquifer which is harmful to humans and biological organisms. Therefore it is imperative that N fertilization should be limited just to increase crop yields but without affecting the environment. In the light of this objective, it is necessary to look into the processes of N uptake and utilization by the plants for growth and grain formation. As the first step it is important to search and identify crop species and cultivars which differ in their ability to absorb N and use them to the maximum in grain formation. Rice, wheat and maize are important crops providing staple food for many millions. The NUE (nitrogen use efficiency) of the cultivars amongst these needs to be examined. The influence of soil properties and soil bacteria has to be included for assessing the NUE of crops (Burger and Jackson, 2004).

Hirel *et al.* (2007) have qualitatively demarcated the N use for various plant processes, like, the growth of the stem, leaves and the flowering and grain or fruit formation. In the first and the earliest process, the root absorbed N is used for the synthesis of amino acids and proteins needed for plant structure and the photosynthetic organs. The second stage relates to the remobilization of the N accumulated and stored, for the flower formation. The proteins are hydrolysed and the amino acids are transported to the fruiting region, the new 'sink'. In wheat (*Triticum aestivum* L.) grain, about 60-95% of N are from the stored N in the roots and shoot (Habash *et al.*, 2006). After the flowering stage, there occurs N deficiency in the plants when grown under N deficiency. In a few crops like wheat, N is given in split doses, based on the total N requirement (Kichey *et al.*, 2007). Some agronomists use the SPAD meter for assessing grain N requirement (Lopex-Bellido *et al.*, 2004). Similarly the chlorophyll meter SPAD is used for N requirement stages in rice (Peng *et al.*, 1996). The selection of the cultivar should be based on low N nutrition, to produce yield not lower than 35-40% (Gallais and Coque, 2005). Hirel *et al.* (2007) have discussed several aspects of higher grain yield with low N input, namely, the mechanisms of N acquisition by roots, NUE in grain composition and grain filling periods, photosynthesis, and plant development.

The NUE has been defined in various ways. Fagaria and Baligar (2003) calculated Agronomic Efficiency (AE) as $(G_f - G_u/N_a) = kg.kg^{-1}$ where G_f is the grain yield of the fertilized plot in kg, G_u is the grain yield in the unfertilized plot in kg, and N_a is the quantity of nutrient applied in kg. They also studied the Physiolgical Efficiency (PE) as $=(Y_f-Y_u/N_f-N_u)=kg.kg^{-1}$ where Y_f is the total biological yield (grain plus straw) of the fertilized plot in Kg, Y_u the total biological yield in the unfertilized plot in kg, N_f is the nutrient accumulation in the fertilized plot in grain and straw in kg, and N_u is the nutrient accumulation in the unfertilized plot in grain and straw in kg. They also proposed Apparent Recovery Efficiency (ARE) as $(N_f-N_u/N_a) \times 100$ as percent. The NUE has been worked out and presented for lowland rice using the formula above. They found that NUE was generally higher for low N rates and decreased at high N rates. Perhaps such analysis could be useful in estimating the likely amount of N needed to be given a particular crop, avoiding excess N application. Recently a new kind of N fertilizers are introduced for crop plants. Controlled release N fertilizers as also the nitrification inhibitors are in use for improving NUE for many crops (Shoji *et al.*, 2001). The former class permits the N uptake by plants according to the demand by the plants and thus reduces the loss of N due to leaching into the ground water. These fertilizers are sulfur-coated urea, polymer-coated water-soluble fertilizer, and less easily soluble and easily degradable materials (Maynard and Lorenz, 1979). The use of nitrification inhibitors has the advantage of improving NUE as well as increase crop yields (Ferguson *et al.*, 2003). The inhibition of nitrification helps retain NH^+ in the case of NH_4^+-N fertilizers and conserve N against leaching in contrast to NO_3-N which is more easily leached down. Nitrification inhibitors also prevents loss of N through N_2O emission in barley (Frye *et al.*, 1989;Delgado and Mosier, 1996).

Rising costs of fertilizers and the environmental harmful effects due to that have heightened the need to develop good fertilizer management methods for crops without affecting the yield (Stevens *et al.*, 2007). The sugarbeet industry is an important economic enterprise in the temperate regions. N management is very important in this crop, since insufficient N reduces the root yield and excessive N reduces the sucrose content. The increased impurities interfere with the sucrose recovery from the roots. Experiments were carried out to compare different preplant N placement methods, in sugarbeet (*Beta vulgaris* L.). N was applied at different rates, using three different strategies, namely, broadcast and incorporated (BI), knife-banded 18 cm from the seed row (KB), or point-injected 8 cm from the seed row (PI). The results showed that PI gave maximum (predicted) yield Y_{max}, with average sucrose yield of 6003-975 kg ha^{-1}, in contrast to BI and KB methods. Furthermore, the N required for maximum sucrose yield ranged from 10 to 100 kg ha^{-1} less for PI than other methods. NUE was the highest for PI, next in KB, and lowest for BI. This effect of PI is considered as due to the placement of N closer to the seedrow, and there was less leaching, with a greater uptake during the early growth periods. PI is concluded as the best method with a greater NUE for sugarbeet.

Biermacher *et al.* (2006) have carried out experiments to develop a precise application system for winter wheat, using a sensor-based one to determine crop N needs. The objective of this study was to determine the expected maximum benefit of a precision N application system for winter wheat that senses and applies N to the growing crop in the spring. Application costs for both dry and liquid sources of N before planting, or during the growing season. They found that the precise system reduced the overall N application by conventional method before planting by 59-82% and the delivery system would cost $ 33 per hectare.

Fertilizers especially the nitrogenous ones applied to soil in excess cause the contamination of ground water, which has become a major issue confronting the environmental scientists. There is a global nitrogen cycle, which starts with the fixation of atmospheric nitrogen, and finally released back into the atmosphere through the burning of fossil fuel. Human activities include mainly the

manufacture of nitrogenous fertilizers. All these result in the acidification, climate change and ozone hole. These and related aspects are discussed by Mansfield *et al.* (1998). Nitrogen is supplied in either as nitrate or ammonium form, and the former being highly soluble, it is rapidly carried down the soil with rainfall or irrigation leaving very little of it in the surface runoff. On the other hand, ammoniacal form is held on to the soil particles only being carried away with erosion and run-off. Conservation tillage which refers to the preservation of soil quality by adopting methods of farming like 'no-tillage' helps prevent the runoff of either form of N. Baker and Laflen (1983) obtained 97% reduction in soil loss for a 'no-till' as against plowing, and a 75 to 90% reduction in total N loss, in two alternate rotations, namely, growing soybeans following corn and 50 to 73% reduction for corn grown after soybeans. Conservation tillage have been effective in reducing N loss in several other crops (Seta *et al.*, 1993). Other conservation practices such as terraces and contouring reduce total N losses due to erosion reduction. But the reduction in nitrate leaching with no-till, differs with the soil type. Experiments in silt loam soils revealed that 'no-till' increased nitrate leaching. Thomas *et al.* (1973) compared the nitrate loss under two treatments, one in growing corn in 'no-till' and the other with conventional system and found that more nitrate was lost from the former treatment which could be direct leaching or due to denitrification. Similarly, nitrate movement under conventional tillage and no-till were compared for 4 years on a silt loam soil in Ontario (Patni *et al.*, 1998) and it was observed that nitrate concentrations in ground water were higher under conventional tillage than no-till every year. The nitrogen leached out from the soil reaches water courses and larger amounts of it in water are harmful to human health. The leaching could be prevented by growing plants on the edges of water front. Riparian buffers, the vegetated region adjacent to streams and wetlands, are thought to be effective at intercepting and reducing nitrogen loads entering water bodies. Riparian buffer width is positively related to the effectiveness of removal of N, by influencing its retention or removal. Studies by Mayer *et al.* (2007) have shown that buffers wider than 50 metres more consistently removed significant portions of nitrogen entering a riparian zone than narrow buffers (0–25 m). Buffers of various vegetation types were equally effective at removing nitrogen but buffers composed of herbaceous and forest/herbaceous vegetation were more effective when wider. Their meta-analysis suggested that buffer width is important for management of N in watersheds. Further, it was found that soil type, subsurface hydrology and biogeochemistry also are important factors governing N removal.

3. PHOSPHORUS FERTILIZATION

Studies on P losses in runoff were made with phosphorus. Logan (1987) found conservation tillage reduced total P losses due to erosion of the insoluble P. Controlled studies have documented the ability of various conservation tillage systems to reduce P losses. In one such study, total P runoff losses were compared between no-till and conventional tillage in Iowa (Baker and Laflen 1983), and with a 97% reduction in erosion with no-till there was 80 to 91% reduction in total P loss for soybeans grown after corn. When corn was grown after soybeans, there was 86% reduction in erosion, and 66 to 77% reduction in P loss.

P is second only to N as the most limiting element for plant growth. The amount of P in plants ranges from 0.05% to 0.30% of total dry weight. The concentration gradient from the soil solution to the plant cell exceeds 2,000-fold, with an average free P of 1 µM in the soil solution (Ragothama, 1999) and is well below the K_m for plant uptake. Thus, although bound P is quite abundant in many soils, it is largely unavailable for uptake (Schactman *et al.*, 1998). Crop yield on 40% of the world's arable land is limited by P availability. P is unavailable because it rapidly forms insoluble complexes with cations and is incorporated into organic matter by microbes. The acid-weathered soils of the tropics and subtropics are P- deficient and high in Al which is toxic to plants (von Uexküll and Mutert, 1995). Bieleski (1973) has estimated that to produce a grain crop yield of 7 metric tons Ha^{-1} needs an addition of 90 to 120 kg P Ha^{1}. However, despite an adequate P fertilization, only 20% or less of that applied is utilized by the 1st year's growth. Runoff of P from the soil causes eutrophication and hypoxia of lakes and marine estuaries in the developed world where excessive P is applied in general. It is estimated that rock phosphate reserves in the soil could be depleted in as little as 60 to 80 years (Runge-Metzger, 1995). Since it is very important for crop production, its use has increased over 4- to 5-folds between 1960 and 2000 and is expected to increase further by 20 Tg $year^{1}$ by 2030 (Vance 2001). Abelson (1999) has forecast that there will a crisis for P fertilizers in the

21st century and plant biologists have a responsibility to discover or introduce a mechanism which will enhance P acquisition or utilization by plant systems, or develop P-efficient germplasm, and advance crop management schemes that increase soil P availability. Vance has advocated the following measures to improve P fertilization. 1. Extensive using of legume crops which would improve symbiotic N fixation, 2. introduce intercropping methods of farming for efficient use of both N and P, 3. orient fundamental research to identify genes which are involved in N and P uptake and utilization and also develop conventional plant breeding technologies to fix these properties in crop cultivars. A recent review suggests various ways for improving the efficiency of plant roots to assimilate the soil phosphorus (Lambers *et al.*, 2006). Unlike N, phosphate is far less mobile and as little as 1-5% of the plant's requirement reaches the plant roots through mass flow and furthermore, the amount intercepted by growing roots is only half of that. The diffusion through soil solution is much slower in the dry soils which contain less moisture. Lambers *et al.* (2006) consider that P uptake can be increased by root interception, and this could be achieved by greater proliferation of the root system, by longer root hairs and also by mycorrhizal symbiosis. The different root traits which are suited for greater P acquisition are also discussed. The increased allocation of root biomass is observed when the P is limited in the soil and this is considered to be ontogenetic in some cases. However, it is also influenced by the amount of cytokinins produced in the roots and exported to the shoot. While discussing the root length for P acquisition, the findings of

Steingrobe (2001) are interesting. It was found that the total root-length production of in the field grown *Beta vulgaris* during the entire growth period was 3-4 times that of the plants at harvest, which was 200 $km.m^{-2}$ and 120 $km.m^{-2}$ respectively at low P supply. As a result, P uptake was 25% more, than in plants at high P supply. Similar increases were obtained in *Hordeum vulgare* (Steingrobe *et al.*, 2001). Increased root turn-over is important for greater acquisition of immobile soil P.

4. CONSERVATION TILLAGE

There has been on-going research in USA to find out if organic farming could be better for increasing the soil properties and also make the crops more productive. The research by the Teasdale (2007) group at ARS, Beltsville, Maryland, showed that 'organic' plots had more carbon and nitrogen and yielded 18% more corn than the other plots with conventional 'no-till' practices. Another study by a group found that in the years of normal precipitation, returns from 'no-till' wheat were similar to those with one crop of winter wheat with a year of fallow in between. Laura McGinnis (2007) writes in the same magazine that "On a smaller scale, economic analysis can also reveal surprising truths about making agriculture profitable". Soil scientists investigated how tillage affects greenhouse gas emissions, when corn leaves a lot of crop residues before tilling the fields for the next crop. The team led by soil scientists found that with adequate fertilizers, the 'no-till' system could conserve more soil carbon that the conventionally tilled soil, and it also reduced greenhouse gas emissions. Another group measured the effects on yield of nitrogen fertilization and tillage which includes both conventional and 'no-till', on irrigated continuous corn. They found that 'no-till' management helped in reducing soil erosion and fossil fuel consumption, sequestering carbon in the soil, and lowering emission of greenhouse gases. But it lowered the yields by about 23 bushels per acre under optimal N dose However, in contrast to earlier findings, further analysis showed that 'no-till' was actually more profitable than the conventional tillage (*see* Figure). Using a statistical model, it was determined that 'no-till' increased net profits through a reduction in the use of machinery, labor costs and fossil fuel consumption.

Conservation tillage practice is a means for increasing profitability and improving soils. While tillage operations like soil preparations, weed control and postharvest stalk management would cost 25% of overall cotton production costs, conservation tillage gives yields compared to the standard tillage practices. Also the reduced-till system decreases the number of tractor operations by over 41%, fuel cost by 48% with an overall production costs by 14 to 18% according the report by Mitchell *et al.* (2006) on their experiments with cotton in San Joaquin Valley. On a nationwide basis about 3.27 million acres (29%) of cotton acreage used conservation tillage or reduced-tillage practices in 2002. From 1992 to 2002, no-till cotton acreage increased 740% in the United States, led by Georgia, North Carolina, Mississippi, Tennessee and Alabama according to a report by Towery (2002). Similarly conservation tillage and cover crops altered favourably the soil properties and increased the soil N and P concentrations. On the other hand, the conservation tillage system caused redistribution of K, loss of organic matter and increased salt concentrations. Cover cropping increased the soil organic matter irrespective of the tillage practice. Veenstra *et al.* (2006) carried out rotation experiments with cotton-tomato for 4 years in San Joaquin Valley and reported their findings.

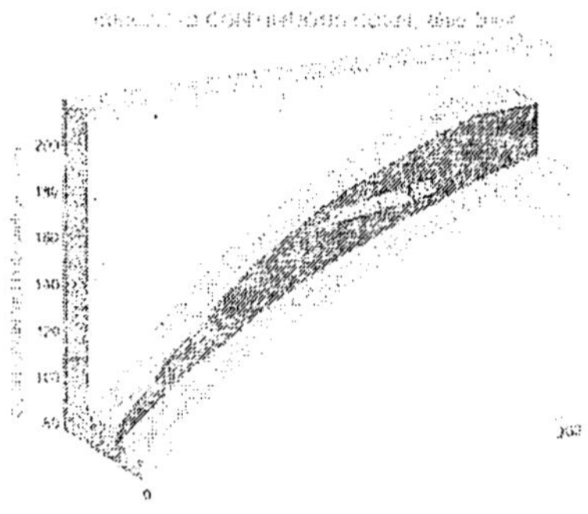

PICTURE 2 (A)

HIGHER YIELDS FROM CONVENTIONAL TILL (ABOVE) COULD MISTAKENLY CONVINCE GROWERS TO AVOID NO-TILL, WHICH IS ACTUALLY MORE PROFITABLE (BELOW)

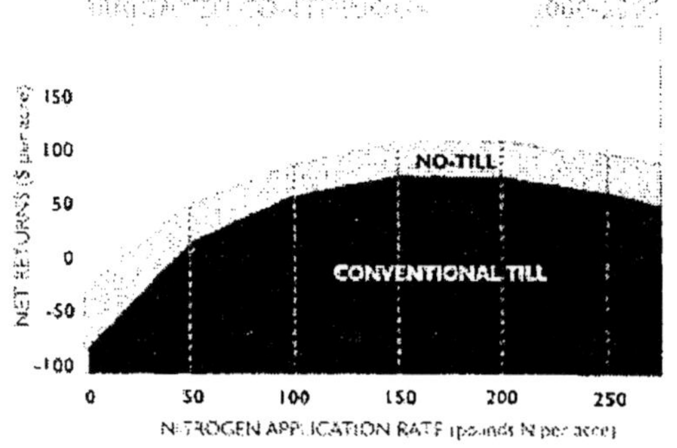

PICTURE 2 (B)

CREDIT: AGRICULTURAL RESEARCH JULY 2007, PAGE 10. *KIND COURTESY*: LAURA MCGINNIS USDA-ARS INFORMATION STAFF, 5601 SUNNYSIDE AVE., BELTSVILLE, MD 20705-5129.

Conservation tillage is a BMP (Best Management Practices) effective in reducing the loss of many surface water contaminants, including sediment, pesticides, and nutrients. However, the effectiveness of conservation tillage depends on factors like the methods and timing of nutrient application, and type of soil, topography, hydrology, and climate. However, some forms of conservation tillage may be less effective in reducing or even increasing nutrient loss. In the case of surface application of manure or fertilizer to no-till fields there may be an increase in the loss of soluble P. Some additional BMPs or changes in tillage practices may be needed in such cases. Many BMPs are available for different crop and livestock production systems that could be adopted to reduce nutrient losses. This subject has been reviewed for major nutrients, N. and P which are of major concern in nutrient losses, and have been discussed by Fawecett (2000).

"Farmers in US who want the best for their fields and families often have to balance ecological and financial concerns" writes McGinnis (2007) in connection with the agricultural system competitiveness and sustainability program of ARS. The objective of the project is to find out the best combination of tools, crops, and management to balance the interest of stewardship and profitability in crop production. The CEAP (Conservation Effects Assessment Project) assesses how US conservation programs are benefiting the water, soil, air and wildlife. Jerry Whittaker, a hydrologist in ARS' Forage Seed and Cereal Research Unit helps develop a modeling system for use in CEAP and other ARS

conservation research. He has made a special software with which one can inform landowners, government agencies, and environmental organizations the benefits of various conservation programs.

Another closely related system is the "Dynamic Cropping Systems" developed by Tanaka *et al.* (2002). The crop production system is adaptive with the development of new technologies in farming research. It has become difficult for the farmers to integrate the vast array of research information which help in sustaining the crop yields under the ever-changing agricultural environment. There are several factors like the weather, input prices, market conditions and government programs which influence the management decisions and one wrong decision could lead to poor economic returns and financial hardship, and also ultimately end to a way of life, as it has happened in Punjab and the states like Maharashtra where farmers committed suicides due to heavy loans. The authors have stressed the need for the farmers to integrate vast amount of information which are constantly changing and this is where the "Dynamic Cropping Systems' help. The key factors of this system are given as (1). *Diversity*, i.e., increasing the type and number of crop species and the variety of products produced, to prevent economic risks. These crop species could be cool and warm season oilseeds, pulse crops, cool and warm season grasses. (2). *Adaptability* which means the willingness to take advantage of new opportunities in cropping practices, like, no-till seeding, better pest control measures and including oilseed and pulse crops. (3). *Reducing the input cost*, i.e., greater net return for the money invested. (4). *Multiple enterprise*—which means including several enterprises to exploit the favorable markets which change frequently. (5). *Environmental awareness*, the farmers must be aware of the environment and natural resources, using high and low residue producing crops for erosion control and soil nutrient management and pest control, and (6). *Information awareness* which is most important, and the producers must make it more competitive, in choosing the crops and varieties and following the BMP (best management practices) as described earlier by Fawecett (2000) under the topic of conservation tillage. Tanaka *et al.* (2002) in their summary, state that this concept of dynamic cropping systems is developed on the premise that farmers/ producers can choose from a variety of crops to optimize or regulate their production by effectively responding to the new situations. For this system the authors have developed an interactive computer information product called Crop Sequence Calculator, available at www.mandan.ars.usda.gov; verified 22 May 2002, in order to help farmers assess crop options and sequencing them in their own cropping systems.

5. LONG TERM FERTILIZER TRIALS

Fertilizers are important for increasing crop production. But this has come under intense scrutiny over the last 3 decades because of the effects on the environment. Food production for the expanding population have led to the intensification of crop management. And this in turn involves a greater need for commercial fertilizer application to prevent nutrient depletion in the soil. There have been greater concerns on the potential negative effects which are more often exaggerated over the actual benefits derived by adequate fertilization which provides abundant food supply affordable by all (Stewart *et al.*, 2005). The authors provide evidence how commercial fertilizers have made significant impact on agricultural production over the last 5 decades. These authors have presented the results of studies made by a number of scientists. The reduction in yield of crops like corn, cotton, rice, barley, sorgum, wheat, soybean and peanut due to withholding of chemical inputs, like inorganic N fertilizers, was analyzed by Smith *et al.* (1990). They estimated that the U.S. corn yield would decline by 41%, 37% in cotton, and 26% in nonleguminous crops, as a result of no supply of N fertilizers. The authors collected information on the experiments described below, from Oklahoma State University (2000) for 'The Magruder Plots', S.J. Troesser, (2003), for 'Sanborn Field', University of Illinois Extension report 2004 for 'The Morrow Plots', and Schlegel, (2000), for 'Long-Term Corn and Grain Sorghum Study', A.E. Johnston, (2003), for the 'Broadbalk Experiment' in England.

There have been long-term fertilizer trials in different places in USA and in England which are carried out over 100 years. 'The Magruder Plots' were established in Stillwater, Oklahoma, in 1892, and are the oldest continuous wheat soil fertility research project in the world. From 1930 inorganic N is the form of sodium nitrate, and from 1946 it was changed to ammonium nitrate applied every year, to supply N at 37 to 67 kg ha^{-1}. Similarly inorganic P was ordinary superphosphate at 9% in the early years, and triple superphosphate at 20% P from 1968. The amount of P was kept constant at 15 kg ha^{-1}. The report over 71 years, showed that N plus P fertilizer applications caused an increased wheat yield by 40%. Girma *et al.* (2007) reviewed the results of experiments on continuous winter wheat (*Triticum aestivum* L) which were initiated by Alexander C. Magruder in 1892 to answer questions of yield response to amendments, yield stability, percentage organic matter (OM) lost over time, soil nutrient status, microbial activity, weed population, and information on the economic returns. The original objective was to evaluate wheat production on native prairie soils without fertilization. After 6 yrs, the original plot was divided into two one of which was fertilized with cattle manure. The experiment was then modified in 1930 by Dr. Horace J. Harper. Since 1947, six treatments were continued for evaluating the effects of simple combinations of manure, and inorganic N, P, K, and lime. For 114 yrs, winter wheat was grown continuously under conventional tillage, with the check plot receiving no fertilizers. Winter wheat production was continued under conventional tillage for 114 years. The check plot which was never fertilized continued to produce grain yields of over 1 Mg ha^{-1}. Despite the decline in soil organic matter from 4 to 1% during this time period in these plots, wheat grain yields actually increased slightly with time instead of a decline. While continuous wheat without rotation is not recommended by the reviewer, this 114-yr study documents the feasibility of growing wheat with no loss in grain yield.

The Sanborn Field trial was initiated in late 1888 for evaluating the effect of rotation and manures on crop yield. Commercial fertilizer was introduced in 1914. The results of N, P and K application on wheat grain yield were analyzed from 1889 to 1998. It was concluded that the nutrients caused an increase of 62% over a 100 year period Similar long-term fertilizer trial is conducted at the University of Illinois and is known as 'Morrow Plots' where crop rotations and fertility treatments were evaluated since 1876. In 1955, commercial fertilizers were included which had urea-N, superphosphate for P, muriate of potash for K and lime. The continuous corn grown in both control and the N,P and K combinations gave 57% increased yield attributed to fertilizer with lime application. When the input was withheld, the loss in yield was very high. Experiments on long-term corn and grain sorghum at Kansas State University farm showed in general, N and P application increased the yield by 44% for corn and 31% for sorghum. The Broadbalk experiment was carried out at Rothamsted Experiment station in England from 1843 on the yield of winter wheat. Their reports showed that N+P+K application increased the yield over 62%. Smith *et al.* (2005) have concluded, "commercial fertilizers make up the majority of nutrient inputs necessary to sustain current crop yields, with available organic sources, native soil reserves, and biological N fixation supplying the remainder". The inorganic fertilizer inputs are essential to produce food as of now, and for the years to come with the increasing population. They are fully convinced that without the current inorganic fertilization, food production cannot be increased, and it will be a great disadvantage if it is reduced for the fear of eutrophication, or contamination of soil and ground water.

Inorganic nutrients are as important for vegetable crops, as for field crops like the cereals. Progress made in this regard is discussed by Hochmuth (2003). Vegetable production accounts for a billion dollar business in the states like California, Florida, Texas and Arizona and a large part of success in vegetable production is due to the research-based understanding of the mineral nutrition of the vegetable crops, especially with regard to fertilization for optimum crop production and quality. Hochmuth states that the establishment of the first state run experiment stations led to the research and demonstration of fertilizer needs for crops. Though in the early periods of the 20th century research was directed towards identification of essential nutrient elements, later studies were concentrated on the optimum fertilizer requirements, and management of application for the production of crop quality without causing environmental hazards. Most of the chemical fertilizers developed were soluble forms and these often resulted in leaching of nutrients like nitrogen into the ground water. Farmers were advised then to use split doses of application. One technology developed later is the controlled-release fertilizers (CRF), or slow-release fertilizers. It consisted of coating the fertilizer material with one that reduced the solubility of the fertilizer pellet. On application, the coating slowly decomposes releasing the nutrient like N, with different periods, which depends on the thickness of coating. One of the earlier compounds was sulphur-coated urea used for vegetables. Further advances led to the coating of the fertilizer materials with water-resistant polymers for N and also mixed nutrient elements which have increased the field vegetable production. Nitrification inhibitors have been introduced in the 1980s, which reduced the conversion of ammonium-N to nitrate-N which is the most leachable forms in the soil. The nitrification inhibitors are nitrapyrin (2-chloro 6-trichloromethyl pyridine) and DCD (dicyandiamide), the commercial use of which have not been widespread in vegetable crops (Frye *et al.*, 1989; Welch *et al.*, 1985). Efforts to control nutrient losses due to leaching in soils were made as early as in the 1950s with the introduction and commercialization of polyethylene sheets as soil covers for vegetables (Emmert, 1957). Plastic mulch was expensive in the early years and researchers introduced the strip-mulch system.

6. PLANT NUTRIOMICS IN CHINA

China has the largest population in the world, about 1.3 billion and the per capita income is said to be less than US $1000 and the population will touch 1.6 billion by the year 2030 quoting the figure from FAO. Yan *et al.* (2006) have made a comprehensive review of the food problems and measures taken to increase food production through innovative and effective solutions. 'Plant nutrionics' is a new frontier area which provides methods for improving the efficiency of plants to utilize the nutrients taken up for crop yield. In China, the traditional farming have been followed over centuries through labor-intensive nutrient management. The 'green revolution' made a great impact on food production in China, and the dwarf wheat and rice varieties which brought the success, were highly responsive to high fertilizer input. China is considered the largest fertilizer consumer in the world, as revealed by certain facts, *viz.*, in 2004, China used 30% of the chemical fertilizers produced in the world, which comprised 20 million tones of N, 10 million tones of P and more than 50% of the cost of crop production is accounted for the cost of fertilizers, according to the figures from China Statistical Bureau. The authors stress the need for a 'second green revolution' which is attainable through evolving crop varieties which will be more productive with less soil fertility. Lian *et al.* (2006) have reported their study on genes responsive to low N stress in 'Minghui 63' *indica* rice. They identified the profiles under low-N stress.

The authors have reviewed the work on phosphorus nutrition. Low P availability is a constraint to crop production in China and the soils are P deficient with a high P-fixing capacity. Most of the cultivated soils are acidic in South China, and calcareous in North China. It is necessary for China to develop a new approach to tackle the P nutrition problem. Crops species which would adapt to low-P state, and those which could be more efficient in uptake and utilization of applied phosphatic fertilizers are to be evolved. Li *et al.* (1995) have reported the identification of a number of P-efficient wheat genotypes. These have been found to secrete more organic acids like malic and citric acids, into the rhizosphere. Chinese researchers are examining the physiological mechanisms of low-P adaptive traits like root morphology, architecture, root exudates, and the spatial configuration of the root system in the soil profile. This is critical for P-uptake efficiency in many leguminous crop species like bean and soybean. Yan *et al.* (2006) also have stressed the

importance of iron nutrition, since iron deficiency is widespread in China.

7. FERTILIZATION AND WATER CONTAMINATION

The use of organic fertilizers is becoming popular because of the perhaps 'misconception' that inorganic chemical fertilizers are harmful for health and these pollute the soil. Though there is some merit in this argument, one thing is certain that organic fertilizers are not available in large quantities to replace inorganic ones, and the latter is nonetheless important for increasing crop yield. Basically, soil organic mater (SOM) is an important source of nutrients for plant growth that needs to be maintained for agricultural sustainability. Conventional farming involves repeated tillage and large use of fertilizers and pesticides which lead to loss of organic matter (OM), and yield reduction. Addition of organic matter is known to improve soil properties and this subject has received much attention in recent times, mainly because of the soil pollution due to excessive inorganic fertilization. Studies comparing the organic and mineral fertilization in soils showed higher SOM and total N for organic-amended soils (Bulluck *et al.*, 2002; Warman, 2005). Usually, the organic-amended soils showed significantly higher soil macronutrient content (Edmeades, 2003). However, other authors reported lower macronutrient content for organically fertilized soils (Gosling and Shepherd, 2005). Some authors showed that the application of organic amendment improved soil nutrient content, but did not always increase plant nutrient concentration (Roe, 1998;Warman, 2005). However, the nutrient content in plants is differentially influenced for different crops, nutrients, and climates (Warman and Havard, 1998). However, there are many definitive findings that organically grown vegetable crops have lower nitrate content (Williams, 2002). Low nitrate content in the edible part of the plants is very important for human health, due to its potential transformation to nitrites, which interact with hemoglobin and respiratory problems (Causeret, 1984). Nitrate concentration in beet and lettuce was very much less in organically raised crops (Bosch *et al.*, (1991). Green manure is a crop grown and used for supplying nutrients, nitrogen in particular, to the crop plants grown subsequently. The nutrients in the organic form in the GM are released to the soil on decomposition. GM approach to crop production improves the soil conditions and reduces the environmental hazard caused by inorganic fertilizers. Unlike synthetic N fertilizers, legumes grown as GM provide a potentially renewable source of on-farm, biologically fixed N and may also fix and add large amounts of C available for the next crop. Though GM-based systems may provide alternatives to current approaches to crop production, GM may not be economically justified without the provision of multiple services such as nutrient supply, pest and weed control, and improvement of soil characteristics for crop production. Provision of such services within GM-based systems requires proper assessment of the interactions between the GM, the environment, and management. Effective farmer participation is imperative for successfully introducing GM based crop cultivation. This is fully discussed by Cherr *et al.* (2006) in the review and interpretation of GM.

Interest in soil organic fertilization has grown appreciably in recent years; however, few studies have been performed in greenhouses. A comparative study of organic vs. mineral fertilization in a greenhouse has been conducted for 9 yr in a calcareous loamy soil classified as Xerofluvent in the Guadalquivir River Valley, Seville, Spain by Herencia *et al.* (2007). The nutrient availability in the soil, macronutrient concentration in the edible part of the plants, and yield were examined. The organic fertilizer used was vegetal compost and green residue of previous crops that came from the experimental farm and did not depend on external inputs. The use of organic fertilizer resulted in higher soil organic matter, soil N content, and available P and K. However few differences were found in the macronutrient concentration in the edible part of the crops, independent of the type of fertilization. The nitrate concentration in the edible parts was significantly lower for the crops grown in the organically fertilized plots. Crop yield was not statistically different between fertilizer treatments. This study demonstrated that long-term use of organic compost in greenhouse soil improved soil fertility and produced similar yields and nutrient composition in the edible portion of crops compared with mineral fertilization.

One way to minimize the contamination of ground water by nitrogen is by protecting the water resources. Don Comis (1999) has described the measures taken by the ARS USDA scientists under the theme, "protecting Water Resources". He writes "Water quality—in terms of contamination by pesticides and nitrogen fertilizer—became an issue in the 1970s.....In 1990, ARS took the lead in an USDA water quality program across the Midwest, studying large areas called Management System Evaluation Areas (MSEA). Since then, the MSEA program has expanded beyond the Midwest, to include the Mississippi Delta and the Eastern Coastal Plains....The MSEA program has resulted in more efficiency in nitrogen fertilizer use and subsequent reductions in nitrate contamination of water.....Farmers in one Iowa watershed study area reduced their nitrogen fertilizer applications by 50 pounds an acre on about 20 percent of the watershed.......A plant chlorophyll meter that tells farmers how much nitrogen a plant needs has proven valuable in guiding nitrogen applications". Comis quotes another scientist, Sharpley, "Over the last 20 years, we have done a good job reducing point sources of pollution with improved water treatment and phosphorus bans in detergent. Now a great share of phosphorus entering our rivers and lakes comes from sources such as agricultural runoff......Phosphorus runoff has always been a concern but we worried more about nitrogen because of the direct health effects of high nitrates in drinking water. Unlike nitrogen, phosphorus binds tightly to soils. We figured if we stopped runoff and erosion, phosphorus would pretty much stay in place....When phosphorus builds up to very high levels in soil, it can wash off as dissolved phosphorus too.....The only permanent solution to reducing soil phosphorus is to balance farm input and output to reduce the amount of phosphorus brought onto a farm as feed and fertilizer, so that crops and animals are fed as closely as possible only the phosphorus they actually need". The "Blue Revolution" is started by the US Dept of Agriculture. Sharpley is working with the USDA Natural Resources Conservation Service for finding the phosphorus thresholds for the various soil types across the US by the year 2002. Comis states that irrigation and drainage are crucial to American agriculture, and the "development of the modern irrigation infrastructure like the center-pivot sprinklers since 1960—has led to a technological evolution in the application of water to the land. What has been termed the "green revolution" in worldwide food production was in many ways a "blue revolution," as irrigation expanded around the world. Today, irrigated agriculture in the US accounts for 40% of total crop value—from only 15% of total cropland".

8. SALINITY

Salinity of the soil poses serious hazard to crop production. The Director of Agricultural Experiment Station of the University of California, Lowell N. Lewis (1984) has made a strong plea to look into the salinity problem which affects soil, "a vital resource in danger", particularly with reference to San Joaquin Valley, California. His editorial runs thus "The wise and efficient management of our natural resources is one of the most demanding tasks facing federal, state and local governments. We have evolved from a land of such plenty that even today it is difficult for some to accept the limited availability and fragile nature of our land, water and energy resources. In a few short decades we have seen an evolution from an economic system that encouraged unrestrained use of these resources, to an era in which we must reckon with finite supplies and devise ways to use our vital resources more judiciously.....Learning to cope with salinization of our soils is not just a job for researchers or farm managers or politicians—it requires the best thinking of all three groups. The stakes are high. Once soil is lost from productivity because of the accumulation of salts, it takes years of treatment, large amounts of water and energy, and careful management to re-establish its viability.....In the rich San Joaquin Valley, with 4.5 million acres under irrigation, more than 400,000 acres are now seriously affected by salinity. It is projected that by the turn the century, in just 16 years, another million acres could be lost to salinity. A loss of that magnitude could mean many millions of dollars in lost agricultural productivity and even greater costs to rehabilitate the land. The nature of the salinity problem in the Imperial Valley is somewhat different, but this region too faces the potential loss of large portions of its productive land.....It will take a coordinated research and extension plan to identify the gaps in our knowledge and educate managers on the best ways to minimize the advance of salinity".

In addition to land, another important resource central to food production is water. Irrigation is closely connected with salinity. If the water is brackish, it adds to the salinity and it is good water which would remove the salts from the surface soil to down below the soil. These salts do come up during evaporation especially in the drier regions, or when there is lack of rainfall, and thus make the surface soil saline. The exploitation of ground water for intensive cropping as in the case of sugarcane which consumes huge amounts of water, also brings up the salts from below and make the soil saline in the course of few years. The data provided by the World Bank show that the spread of irrigation contributed between 50 and 60% of the massive increase in agricultural output of the developing countries from 1960 to 1980. However, further expansion of the irrigation development will be more expensive than it was in the past, because the low-cost water exploitation has been completed already. In some developing countries, the supply of water for irrigation faces another problem which is the silting up of reservoirs. In some areas due to deforestation and overgrazing followed by large scale erosion, silting up of reservoirs takes place more rapidly. The efficient gravity-flow systems made up of basins and furrows which are followed in recent times, can be improved by using lasers to guide machines that level the fields and make it flood them quickly and uniformly. Another system, the trickle or drip irrigation is superior to gravity-fed system. Trickle system is widely used in Israel and US and in other countries too. This system increases efficiency of use of irrigation water and also prevents salinization.

9. SOIL CONSERVATION

Soil conservation in the Philippines

Soil conservation measures are advocated and practiced in many countries where soil erosion is a serious threat to farming. In the Philippines, soil conservation practices are effectively followed for crop cultivation in slopelands, where soil erosion takes place easily. More than half of the Philippines' total land area of 30 million hectares has slopes steeper than 18% (Escano and Tabab, 1998). Out of the total 8 million hectares of arable land, 58% is eroded to some degree, and 41% of farming activity is carried out in this steep hilly land. The productivity of the slope-lands is declining alarmingly due to the soil erosion. Several soil conservation measures have been developed during the past several years using technologies developed by research. The development of SALT *(Sloping Agricultural and Land Technology)* was developed in 1971. SALT 1 advocates alley cropping using leguminous tree or shrub species planted closely in a belt along the contour lines. A mixture of food and cash crops are planted between the rows. SALT 2 recommends land use of 40% for agriculture, 20% for forestry and 40% for livestock like goats. Forage crops, and cash and food crops are planted as mixture. SALT 3 is the sustainable agro-forest land technology where forest trees are planted in slopes of more than 50%. Tree species are a mixture of fruit and timber crops. SALT 4 recommends planting of fruit trees on the upper two-third portion of the farm.

Soil Conservation in Taiwan

The island of Taiwan is 142 km broad at its widest point and 377 km long.

The central mountain range runs along the length of the island surrounded by a narrow range of foothills. From the foothills, the terrain gradually slopes to the land 100-500 meters in elevation. In response to economic pressure, much of Taiwan's agriculture has migrated into the high slope land areas. The resulting increase in deforestation and land disturbance has increased the soil erosion. The climate in Taiwan is subtropical in the north and tropical in the south. Rainfall is plenty, and concentrated during the summer and winter. The heavy rainfall causes landslides and flooding and attendant soil erosion. The government established soil conservation project in 1952 and a number of practices have been developed. These include hillside ditches, bench terraces, grass barriers, cover crops and mulching (Chia-Chun Wu, 1998).

Soil Conservation on Sloping Orchards in Japan

Japan is located in the Asian monsoon zone and has a rainfall of 1000 to 3000 millimeter per year. More than 50% of the rainfall occurs in summer. Since about 70% of the land is hilly or mountainous, it has a long history of development and improvements for slope land agriculture. In Japan, most orchards are located on sloping land. Citrus orchards especially are established on steep slopes. Soil conservation viz., mulching and planting of grass cover are widely used to prevent erosion. In addition engineering measures such as the construction of ditches are resorted to (Seiji Nakao, 1998).

Soil Conservation in USA

Soil conservation measures are followed in US and it is considered that a good crop can be obtained using minimum tillage—even just enough to plant seeds. Comis (1999) states that "tillage tends to have an adverse effect on soil health. Good soil has more organic matter, is less erodible, cycles nutrients more efficiently, and holds more water.....ARS researchers over the years, working with NRCS and universities, have done the supporting research that encouraged farmers to largely put away the moldboard plow—used in breaking out the American

West—and reduce erosion by 90% on about 40% of US planted acres....We now hypothesize that this dramatic change in tillage techniques may have shifted US farm soils from net CO_2 producers to net accumulators of carbon—in the form of valuable soil organic matter. This would make these soils more productive and part of the potential global warming solution, rather than part of the problem."

10. SOIL EROSION IN KENYA

In 2005, Professor Rob Dunbar and his colleagues collected core samples from a large coral colony near Easter Island. Kenya. "Coral reefs, like tree rings, are natural archives of climate change. But oceanic corals also provide a faithful account of how people make use of land through history, says Robert B. Dunbar of Stanford University". In a study published in the Feb. 22 issue of Geophysical Research Letters, Dunbar and his colleagues used coral samples from the Indian Ocean to create a 300-year record of soil erosion in Kenya, the longest land-use archive ever obtained in corals. A chemical analysis of the corals revealed that Kenya has been losing valuable topsoil since the early 1900s, when British settlers began farming the region. "We found that soil erosion in Kenya increased dramatically after World War I, coinciding with British colonialism and a series of large-scale agricultural experiments that provoked a dramatic change in human use of the landscape....Today, the Kenyan landscape continues to lose topsoil to the Indian Ocean, primarily because of human pressure", said Dunbar. Erosion is a serious threat, he noted, because the loss of fertile soil often is accompanied by a decrease in food production. According to one recent study, soil erosion is a global problem that has caused widespread damage to agriculture and animal husbandry, placing about 2.6 billion people at risk of famine. "This is particularly worrisome in East and sub-Saharan Africa, where per capita food production has declined for the last half-century," Dunbar said (see Shwartz, 2007).

11. ISRAEL'S ECOLOGICAL TRIUMPH

Another aspect closely related to the soil degradation, is desertification. With the continuous exploitation of soil and water, land becomes less arable leading to desertification. The story of combating the desertification and "Making the Desert Bloom" has been described by Tal (2006). There was the first academic conference in Israel focusing the attention of the participants on "Deserts and Desertification"-Challenges and Opportunities". Desertification, or the degradation of soils in the drylands, is a global problem and are of serious concerns to the countries all over. According to a World Bank estimate, 10-20% of the world's drylands come under severe degradation and over 400 million of the world's poorest people are affected by this kind of degradation. How the reclamation of the desert land was done is a great story. Tal has described that "Even with its meager resources as a nascent developing country, Israel set about to reclaim its desert heritage. During the 1950's, infrastructure projects were built that delivered water from the rainy north to the desiccated south. Settlements were established that first invented and then expanded drip irrigation technologies to produce prosperous, local, agricultural economies. The Jewish National Fund succeeded in planting trees on dry and salty lands that professional forestry literature had long since written off. Grazing was organized, with seasonal allocations made to ensure that the land's carrying capacity was not exceeded". The plenary session ended with a call for greater commitment to assisting the developing world's dryland nations in their ongoing efforts to protect their soil offering a message to Israelis and to the participating 35 countries that "with ingenuity, collaboration and respect for ecological constraints, the dry lands can be a source of economic opportunity and spiritual edification". In the conference announcement was made for the establishment of an international network of desert research stations initially from the Blaustein Institutes for Desert Research.

12. MICRONUTRIENTS IN CROP PRODUCTION

The micronutrient elements which are essential for crops are Fe, Mn, Zn, B, Cu, Mo, and recently Ni, Co and Si have been added to the list. Micronutrient deficiencies in crop plants have increased in recent times and the reasons are varied, ranging from intensive cropping, increased application of fertilizers containing the major nutrients, and extension of farming to the lands which are of marginal productivity. There are several measures to supply the micronutrients for correction of deficiencies. A good review has been presented by Fageria *et al.* (2002). Plants' acquisition of micronutrients from the soil is influenced by soil conditions and the microbial and environmental factors which help release them to the plant roots. Soil moisture, pH, and soil organic matter also affect the availability of micronutrients. Furthermore, plant factors like the root structure, root hair morphology, rhizosphere induced changes in the release of organic acids and compounds known as '*phytosiderophores*' control the micronutrient supply. Extensive studies by Brown group in the USDA Beltsville laboratory in the 70s have revealed the presence of nutrient tolerance and susceptibility mechanisms, with regard to Fe in particular, and their identification of crop cultivars made a new beginning in nutrition. In the last 3 decades, considerable advances have been made in this field. Chemicals released from the roots like the *phytosiderophores* have been identified. A detailed discussion on the physiological genetics has been made in the recent book by Epstein and Bloom (2005).

13. FOLIAR APPLICATION OF PLANT NUTRIENTS

Application of fertilizers to the soil has decided advantage for field crops and has been practiced over a long time. However, the disadvantages are the overfetilization and the resulting eutrophication of soil and ground water. Slow release fertilizers have been developed to get over this problem. The fact that the nutrients could be supplied through foliar sprays has been known for over 6 decades and is practiced for a long time for field crops like sugarcane and pineapple.

The effects of foliar application of major nutrients like N, P and K on crop yield were carried out in the 1950s at Rothamsted, and the rates of foliar absorption and translocation of several nutrients in vegetables and fruit trees were investigated at East Lansing, and at Cornell.

There have been considerable interest in foliar uptake and transport of nutrients in several plants. The mechanisms of absorption, transport through the cuticular membrane covering the leaf surface have been studied in the last 5 decades. Some recent studies on the effectiveness of foliar supply of major nutrients and micronutrients will be discussed here.

Improvement of the grain protein in high yielding cereals has been the most important goal in recent times because of the high premium it fetched the farmers, and studies have clearly shown that foliar sprays of N increased in grain protein in wheat. Bly and Woodard (2003) examined the optimal timing for N sprays in wheat and found that post-pollination foliar N gave the highest grain protein. Woolfolk (2002) found significant increase in total grain N and protein content from the post-flowering sprays of urea-ammonium nitrate (UAN) in winter wheat. Similar increases on grain protein content from foliar application of N near the flowering stage in wheat has been obtained (Wuest and Cassman, 1992).

Several interesting findings have been reported by Peuke *et al.* (1998) on the absorption and utilization of N and its influence on the carbon metabolism in *Ricinus communis* L. N from the $(NH_4)_2SO_4$ was taken up more readily than that from KNO_3, and more evenly distributed between the shoot and the root while N from nitrate was transported essentially to the roots. Furthermore, the loss of carbon due to dark respiration in the shoot was high following the sprays. The spraying of N in either form increased organic or inorganic particles on the surface of the leaf increasing its wettability and thus increasing the rate of absorption.

Foliar sprays of K salts have been promising. K requirement has increased recently in many regions because of periodic drought conditions resulting in soil compaction and poor soil availability. Soil K availability is very much reduced in Missouri, Illinois and Kansas states, where the subsoils are highly clayey. Lesser amounts of K was applied as fertilizer, by the farmers due to low commodity prices. Nelson *et al.* (2005) studied the response to foliar-applied K at several growth stages and the cost-effectiveness, for soybean in claypan soils. K as K_2SO_4 was given as a pre-plant soil at 140, 280, and 560 Kg ha^{-1}, or foliar supply at 9, 18, and 36 Kg ha^{-1} at V4, Rl-R2, and R3-R4 stages of development. The results showed significant increase in grain yield from 727 to 834 Kg ha^{-1} when K spray was given at 36 Kg ha^{-1} at the V4 and R1-R2 stage, but when given at the R3-R4 stage, grain yield increased but not as high as at V4 or R1-R2. It was also revealed that foliar K could only be a supplemental option when the climatic and soil conditions are unfavorable to root uptake.

Foliar application of K have been used as a supplement to soil application in a few crops. In Arkansas, research showed that the K requirement of the fast-fruiting and high yielding cultivars of cotton far exceeded the uptake capacity of the plants from the soil. Furthermore, the root activity of these cultivars actually decreases during flowering and boll development, when there is an increasing demand for K. Howard *et al.* (1998)

evaluated various K compounds viz., KNO_3, K_2SO_4, $K_2S_2O_3$ and KCl, and buffers for the spray solution, including addition of B.

There are reports that foliar application of nutrients, increased both crop yield and quality. Haq and Mallarino (2005) studied the effects of foliar application of N-P-K mixtures with or without S, B, Fe and Zn at V5-V8 stages on the oil and protein concentrations in soybean (*Glycine max* L. Merr). The results of showed that foliar fertilization increased the oil concentration in one trial and protein in another trial, but decreased the protein content in 2 trials. The effects of P fertilization were not consistent. They concluded that foliar fertilization would not offset the fertilization costs, unless the cost of application was reduced by including it with herbicide.

Calcium sprays are very effective on improving the quality of fruits. Kadir (2004) studied the response of 'Jonathan' apple trees (*Malus dmestica*, Borkh) grafted on EMLA 111 to $CaCl_2$. Trees were given 1 to 8 sprays at 8.971 kg ha^{-1}. More than six sprays improved fruit quality, increase in fruit weight, size, appearance, redness and less scald incidents. The ratio of soluble solids to titratable acidity was also increased by frequent Ca sprays. Fruit skin redness was the most significant effect. This increase was the result of a linear increase in amount of Ca in fruit and leaf tissues and also was reflected in the increase in K, Mg, P and N. Thus the most beneficial effect of Ca sprays was the fruit quality brought about by changes in mineral balance.

Effectiveness of foliar application on different crops as reported by several workers (summary. See Kannan 1990 for details).

NITROGEN	
Urea, uramon, $CaNO_3$, Amm sulphate	sugarbeet, apple, tomato, peach, grape, onion, sugar Beet, potato
Uramon	celery
PHOSPHORUS	
Phosphate or orthophosphate peanut.	sugarbeet, onion, beans, cabbage,
Ca glycerophosphate	Tomato, grapes, apple
Diammonium phosphate	
CALCIUM	
$CaNO_3/CaCl_2$	greenhouse Plants
2% $CaNO_3$	"Topple" disease in gladiolus
$CaCl_2$	"blossom end rot" in Tomato

Iron is one of important micronutrients and foliar sprays are generally effective in correcting chlorosis. Iron chelates have found increasing usage for sprays. In a recent study, the interaction of FeHEDTA given along with and three post-emergence broadleaf herbicides was examined (Franzen *et al.*, 2003). Three herbicides, *acifluorfen, imazamox* and *lactofen* were applied with or without FeHEDTA. At one location, Fe amendment lowered the yields with *acifluorfen* and *lactofen*. But yields were higher with Fe in the *imazamox* conbination, although weed control was less effective. They concluded that Fe combination with the 3 herbicides would not be recommended for correcting chlorosis, and FeHEDTA perhaps should be given alone.

14. AGRICULTURAL REVOLUTIONS AND FOOD PRODUCTION

Writing about the "The Next Agricultural Revolution", Wallace (1984) reviewed the achievements through the Green Revolution and the limitations for yield increase of crops in the years later with the increasing mouths to feed. The GR was a huge success on a worldwide basis although there were some disappointments. It resulted in large yield increases in basic food grain crops like wheat, rice and corn, and the introduction of double and triple cropping where only one crop a year was in vogue. Nations like Japan and India, which imported grains, turned into grain exporters. World hunger subsided with the larger number of third world nations adapting to the technologies of the GR. With the increase in population at the beginning of the new millennium, many leaders in agriculture prophecy the advent of another revolution, viz., Genetic Engineering Revolution in the next 2 decades after the 1980s. Stelly (1980) recognized the difficulties and called for mobilizing the resources so that the results could be obtained at least after 2 decades. The objective is to develop high yielding new varieties of crop plants with resistance to the environmental and nutritional problems as well as pests and diseases. It is further hoped to achieve break-through in three important biological processes. The most important of these are: (i). to transmit the gene for inducing symbiotic nitrogen fixation present in legumes like alfalfa to other non-symbiotic crop species, (ii). Introduce the efficient C_4 mechanism of photosynthesis enjoyed by corn and sugarcane, for example, to other crops, (iii). Develop methods to inhibit the dark photorespiration process so that loss of carbon fixed through photosynthesis could be prevented or minimized in some crops.

Wallace strongly advocates adoption of the Multidisciplinary Revolution. Agricultural science, which is divided into a number of disciplines, should be brought together to work for a common goal. Some disciplines devoted for crop production concentrate on crop selection and management, land reclamation, soil fertility, weed control, irrigation, plant physiology, entomology, plant pathology and so on. Each of these has developed procedures for improving crop yields, but these generally do not involve all of the disciplines to achieve the maximum potential. Putting all together into a unified or integrated system is the basis of the multidisciplinary revolution. In recent times, a number of new words has been coined about 'revolutions' in order to gain publicity, and one such is the 'evergreen revolution'. No revolution in agriculture can be successful without giving importance to the role of plant inorganic nutrients which are as vital for crop production as 'water'.

Reviewing the approach, Wallace reports that in England a whole new concept of growing wheat has been developed. Timely use of fertilizers, fungicides and growth regulators caused yields of ten to eleven metric tons per hectare while the world average was only 3.5 MT per ha for non dry-farm wheat, and that in the US it averaged only 2.5 MT per ha. Interestingly, the management system in England is sophisticated. The growth regulator, *chloramquat* is applied two to three times to a crop to regulate the balance between shoot and root growth. N, P and K nutrients are never allowed to be limiting. Measures to control diseases were taken up. In Washington State where average potato yields are around 45 MT per ha, Dr. Kunkel controlled 11 environmental and management factors simultaneously and the yields were boosted to 100 MT per ha (Anonymous 1979). In USA soybean yields average around 2 MT/ha; however many good farmers get 4-4.5 MT and even up to 7 MT. Therefore application of present know-how can give the high yields obtained by good farmers and researchers, but no one part of the know-how could be omitted to achieve the goal. Wittwer (1975) has shown that record yields to average yields give a ratio of about 3 to 7. Wallace has found that by overcoming several limiting factors to crop production simultaneously, there is an additive effect on the yield. Correcting two factors alone results in a 20% yield increase for each of the factor and when six, eight, ten or more limiting factors are corrected simultaneously, the total effect on yield would be staggering. Hussain (1982) has reported that an integrated approach to overcome limiting factors on sugarcane growth increased yield of the cane almost five times. Wallace (1990) emphasizes the need to eliminate the limiting factors due to soil and plant nutrients, along with others, if the goal to maximize the yield and overcome the yield barrier has to be achieved. He has discussed the nature of interacting factors in detail. According to him, if maximum yield is not attained, it is because of the action of additive or synergistic accumulation of all stresses. If yields are increased on the other hand, it is because of the removal or overcoming of the stresses. As an example, if four different nutrients are present in supply and ratio each at 90% of optimum while all other nutrients are optimum, the summation of the nutrient stress is 0.90 x 0.90 x 0.90 x 0.90 = 66%. Then this yield is possible to obtain, if the interactions are sequentially additive. Small departures from optimum, therefore of several nutrients simultaneously would present serious barrier to maximum yields. A goal of new nutrient efficient cultivars should be to allow high return per unit of input. Furthermore, in the development of new cultivars, it is important that a complete package for nutrient use efficiency be developed, so that the high yields without loss of crop quality are obtained.

Attempts should be made to raise the yield levels on the existing agricultural lands. To meet the food demands, the average yield of all cereals must be increased by 80%. Furthermore, using the currently available technologies, yields can be doubled in the Indian subcontinent, Latin America, and East European countries, and by 100-200% in sub-Saharan Africa, if three conditions, viz., political stability, entrepreneurial freedom, and all the production inputs are guaranteed. More food can be produced if new lands, i.e., lands that are hitherto considered unfit for supporting crops, are brought under plough.

Wallace and Wallace (2003) have discussed various methods for increasing crop production, and introduced the concept of "The Law of the Maximum" for closing the crop-yield gap. They have extrapolated both the concepts of the Liebig's Law of the Minimum and Mitscherlich's law of the Minimum, and introduced the concept of the law of maximum. "The greatest response for any one input is obtained when very few limits remain. Input factors interact to multiply the value of the other when other limiting factors are corrected....Limiting factors are multiple, and their interactions, which can be graphed to give Multiple Action Yield Fraction plots, determine final yields. The MAYF may be used to maximize specific aspects of crop production, such as per unit of land or per unit of irrigation water". They have listed salient points needed for consideration to close the gap to achieve high yields:

Better soil and better management of agronomy are important. The latter would imply the introduction of new cultivars through genetic engineering. In diagnostic programs, complete analysis of essential and nonessential elements is important. Plant breeding should include nutrient efficiency of the newly bred cultivars. Sixty or more factors could be limiting the maximization efforts. "A 50% or even greater gain in yield per acre of many crops can be expected in the next 20-40 years with the use of the principles outlined in the Law of the Maximum. More production on a smaller land base could release considerable marginal land, which could be returned to conservation." Though this was highlighted nearly 23 years ago by Wallace (1984), no importance to this concept has so far been given anywhere. This is reiterated in their recent book (Wallace and Wallace, 2003).

15. GREEN REVOLUTION THEN AND NOW

Norman Borlaug who received Nobel Peace Prize in 1970 for his work in developing high yielding wheat at CIMMYT, and 30 years after that, still is hopeful that there would be enough food for the billions, if biotechnology along with advancements in other fields are properly employed. He considers that the green revolution is a temporary victory in our efforts to fight hunger. He at the same time warns that unless the frightening power of human reproduction is curbed, the success of human efforts to feed the population will be only ephemeral. During the 1990's alone, the world's population grew by a billion people, and again would grow by another billion in the first decade of the 21st century, and will touch 8.3 billion by 2025, stabilizing at 10 billion at the beginning of the next century. He has sounded one important note regarding the use of organic fertilizers. *While the rich and affluent nations who do have enough and more food, can take to "organic" farming, it is no solution for the chronically undernourished one billion people, with low income in the countries facing chronic food shortage.* The need is urgent and the quantitative requirement is high. Furthermore, there is not enough "organic fertilizer" needed for food production for today's six million population. If for example, it is attempted to produce the equivalent of the 80 million tonnes of nitrogen from manure, for such a task, world cattle production have to be increased to five to six billion heads, which is impossible. It is important that scientific education should be imparted at all levels, so that the needless confrontation of consumers against the use of transgenic crop technology can be avoided. Today's bread wheat is an example of a plant genetically modified by nature itself perhaps about 3500 B.C. In the last few years, attempts are being made to use new and appropriate technologies to raise cereal crop yields. Currently large commercial areas are planted with transgenic varieties and hybrids of cotton, maize and potato that contain genes from *Bacillus thuringiensis*, which effectively control several insect pests. There are also many crop plants which have tolerance to a number of herbicides. Thus the herbicides could be used in small doses.

Writing in the editorial in Agricultural Research journal, Welch (1999) describes "For centuries, the success of agriculture has been measured by the number of bushels harvested per acre. ...food producers in the future will have to do more than put up big yield numbers to meet the nutritional needs of the world's populations". People not only need to have enough calories but also enough nutrients including micronutrients. The Green Revolution resulted in increased overall production of grains supplying carbohydrates and proteins, the diet of more than a 2 billion people consume diets that are less diverse than they were 30 years ago which has resulted in over-dependence on a shrinking number of high-yielding staple foods. Furthermore, a growing number of people have diet with inadequate levels of micronutrients, vitamin A, iodine, iron, selenium and zinc. The elements are as important for humans as for plants. The consequences of micronutrient deficiency are enormous in terms of health-care costs. Deficiencies of iron, iodine and zinc cause mortality and decreased cognitive abilities in children born to deficient mothers and so on. The ARS is looking at the human nutrition problem, and among the many, the US Plant, Soil and Nutrition Laboratory at Ithaca, New York is one of the few laboratories in the world whose mission is to improve human and animal health through research on nutrient movement throughout the soil-plant-animal food chain. Welch emphasizes that there should be a new approach to agriculture. He states, "With 2 billion people suffering from poor nutrition around the globe, nations must produce not only more food, but also more nutritious food. This will require a change in how we think about agriculture. A holistic approach to food production could hasten Third World development by improving human health and well being, helping to bring about a 'greener' revolution". Breeding staple plant foods for higher iron, zinc, and provitamin -A (carotenoid) concentrations could contribute greatly to improving the micronutrient status of the diet.

A look back about 50 years from now, the state of agriculture in India particularly in the present Tamil Nadu is quite revealing. It was 5 years of the Independence years, and the end of the World War II. The government in the State and that at the Center gave great importance to agriculture. During the world war time, food grains were supplied through rations and the landowners and tillers were free to use their produce for their family consumption, a situation which continued for several years later. The admissions to agricultural colleges were doubled and great thrust was given to the agricultural extension work, which is carried out by agricultural graduates. Traditional farming included cultivation of rice varieties, which were high yielding selections. Many of these like ADT 8, GEB 24 and popularly known *Chingleput Sirumani* were grown for their quality grain and fairly high yields. However, some of these were susceptible to rice blast disease. Nitrogen and phosphate fertilizers were in use however sparingly and by farmers who were well-informed about them. Generally, the cattle manure available in plenty were extensively used, and supplemented by cattle-penning, and sheep-penning. The cows, bullocks, and sheep were allowed to stay overnight in the fields so that their excreta could improve the soil fertility. In addition, the tank silt was dredged out and applied to soil during the off-season, i.e., just one or two months before the sowing and transplanting of rice. Green-manuring and green leaf manuring were extensively practiced. A decade and more after independence plowing with country wooden plow was replaced by mould-board plow and then with the advent of tractors, animal population mainly bullocks dwindled dramatically. At the same time, different kinds of fertilizers were manufactured in the country with the establishment of Fertilizer Corporation of India in many places in the country.

High-yielding fertilizer responsive ones introduced from the International Rice Research Institute at Philippines, and also by those, which were evolved indigenously, using the IRRI parents has replaced most of the rice cultivars. Thus the traditional farming of the 1950s has changed to unprecedented levels. However, there have been disastrous effects due to over-fertilization.

16. CHALLENGES, CONSTRAINTS, NEW FRONTIERS

There are several constraints, which limit crop growth in certain types of soils. The vast acid-soils found in Brazilian *cerrado*, *Ilanos* of Colombia and Venezuela, Central and South Africa, Indonesia, Thailand and Vietnam, and some regions of India cover well over 8 billion acres of otherwise arable land and are at present not productive. These soils contain high levels of aluminium which is toxic to plants. Kochian a scientist with the USDA-ARS Plant, Soil and Nutrition Laboratory, Cornell University, has been working on finding ways to grow crops on marginal lands such as acid soils. The element Aluminium is the third most abundant element in soils. At neutral or alkaline soils however, it does not cause problem. But in acid soils, Al^{+3} the oxidized form gets solubilized and turn toxic to plant roots. Kochian and his team are investigating the mechanisms of Al tolerance in plants like *Arabidopsis* and have identified some mutants of *Arabidopsis sp.* According to him, the ultimate goal is to isolate the genes conferring Al tolerance, and then introduce this trait in Al sensitive crop species. This study is important because crop production can be increased when the acid soils could be cultivated with suitable adaptive crops. The program developed by Kochian group is named "Phytoremediation", a method for using plant to clean up soil. Besides Al, there are heavy metals like cadmium, cesium, and zinc. Such elements get into food chain through plant uptake, and cause human health problems.

Phytoremediation is the use of green plants to remove such pollutants from the environment or render them harmless. Certain plant species—known as metal hyper accumulators—have the ability to absorb these from the soil and accumulate them in the easily harvested plant stems, and leaves, which can be collected and destroyed. Thalaspi a small weed and a member of the cabbage family thrives on soils with high levels of zinc and cadmium. It can accumulate up to 30,000 ppm of zinc and 1,500 ppm of cadmium in its shoot, without any toxic symptoms, while a normal crop plant is affected by as little as 1,000 ppm of zinc or 20-50 ppm of cadmium in its shoots (Kochian, 2000). There are also plants like kenof and canola which are accumulators of selenium, which is however an essential element in the nutrition of livestock.

Due to intensive cropping, mineral excesses and deficiencies occur extensively in many soils. One of the serious problems especially in alkaline soils is caused by Fe deficiency. Fe deficiency is due to the presence of Fe in the oxidized ferric form instead of the more soluble ferrous form. However, there are certain crop cultivars which have developed mechanisms to absorb Fe from the ferric form. In graminaceous monocot plants, the roots secrete mugineic acid which helps in the acquisition of Fe. Another compound, Nicotianamine (NA) is an indispensable compound in the Fe acquisition mechanisms in Strategy II plants (For more details *see* Marschner, 1995). NA is produced in all the plants and is present in many plant organs. Under Fe-deficient conditions, Strategy II plants secrete mugineic acid family phytosiderophores (MAs) which are natural Fe chelators, which solubilize Fe. NA is an intermediate in the biosynthesis of MA. The group led by Satoshi Mori is working on the cloning of NA synthetase genes (Suzuki,*et al.*,1999) which may ultimately lead to the identification and transfer of genes to plants which are Fe-inefficient.

REFERENCES

Abelson, P.H. 1999. A potential phosphate crisis. Science 283: 2015

Anonymous, Better Crops with Plant Food, Stepping up the yield ladder, a special issue 1979.

Baker, J.L. and J.M. Laflen. 1983. Water quality consequences of conservation tillage. J. Soil and Water Cons. 38:186-193.

Bieleski, R.L. 1973. Phosphate pools, phosphate transport, and phosphate availability. Annu Rev Plant Physiol 24: 225-252

Biermacher, J.T., F. M. Epplin , B. W. Brorsen, J. B. Solie and W. R. Raun. 2006. Maximum benefit of a precise nitrogen application system for wheat

Precision Agric. 7:1-12.

Bly, A.G., H.J. Woodard. 2003. Foliar nitrogen application timing influence on grain yield and protein concentration of Hard Red Winter and Spring wheat. Agron. J. 95: 335-338.

Bosch, N., J.R. Martý´nez, and M.L. Pe rez. 1991. Influencia del tipo de ábono sobre acumulacio´n de nitratos en vegetales. An. Bromatol. XLIII-2:215–220.

Britto, D.T. and H.J. Kronzucker. 2004. Bioengineering nitrogen acquisition in rice: can novel initiatives in rice genomics and physiology contribute to global food security? BioEssays 26:683-692.

Bulluck, L.R., M. Brosius, G.K. Evanylo, and J.B. Ristaino. 2002. Organic and synthetic amendments influence soil microbial, physical and chemical properties on organic and conventional farms. Appl. Soil Ecol. 19:147–160.

Burger M, Jackson LE. 2004. Plant and microbial nitrogen use and turnover: rapid conversion of nitrate to ammonium in soil with roots. Plant and Soil 266, 289–301.

Causeret, J. 1984. Nitrates, nitrites, nitrosamines: Apports alimentaires et sante´. Ann. Fals. Exp. Chim. 77:131–151.

Cherr, C.M., J.M.S. Scholberg and R. McSorley. 2006. Green Manure Approaches to Crop Production: A Synthesis. Agron. J. 98:302–319.

Comis, Don, Farming and Natural Resources, Agricultural Research 1999, Dec., pp 4-9.

Crosson, P. R., and N. J. Rosenberg, Strategies for Agriculture, Sci. American, 1989, 260:128-135.

Delgado, M.J. and A.R. Mosier. 1996. Mitigation alternatives to decrease nitrous oxides emissions and urea-nitrogen loss and their effect on methane flux. J Environ Qual 25:1105-1111.

Edmeades, D.C. 2003. The long-term effects of manures and fertilizers on soil productivity and quality: A review. Nutr. Cycling Agroecosyst. 66:165–180.

Emmert, E.M. 1957. Black polyethylene for mulch vegetables. Proc Amer Soc Hort Sci 69:464-469.

Epstein, E. and A. J. Bloom. 2005. Mineral Nutrition of Plants: Principles and Perspectives. Sinauer Associates Inc. Publishers, Sunderland, Mass.

Escano, C. R., and Sonny P. Tabab.1998. Fruit production and the management of slopelands in the Philippines, Food & Fertilizer Technology Center Extension Bull 450, 1998.

Fageria, N.K. and V.C. Baligar. 2003. Methodology for evaluation of lowland rice genotypes for nitrogen use efficiency. J Plant Nutr 26:1315-1333.

Fageria, N.K., V.C. Baligar and R.B. Clark. 2002. Micronutrients in crop production. Adv Agron 77:185-268.

Fawecett, R. 2000. A Review of BMPs for Managing Crop Nutrients and Conservation Tillage to Improve Water Quality. (Edited and Updated by Tim Smith). Conservation Technology Information Center. West Lafayette, Indiana, USA

Ferguson, R.B., R.M. Lark and G.P. Slater. 2003. Approaches to management zone definition for use of nitrification inhibitors. Soil Sci Soc Amer J 67:937-947.

Franzen, D.W., J.H. O'Barr, R.K. Zollinger. 2003. Interaction of a foliar application of iron HEDTA and three post-emergence broadleaf herbicides with soybeans stress from chlorosis. J. Plant Nutr. 12: 2365-2374.

Frye, W.W., D.A. Graetz, S.J. Locascio, D.W. Reeves and J.T. Touchton. 1989. Dicyandiamide as a nitrification inhibitor in crop production in the southeastern USA. Comm Soil Sci Plant Anal 20:19659-1999.

Gallais A, Coque M. 2005. Genetic variation and selection for nitrogen use efficiency in maize: a synthesis. Maydica 50, 531–537.

Girma,K., S. L. Holtz, D. B. Arnall, B. S. Tubaña and W.R. Raun. 2007.

The Magruder Plots: Untangling the Puzzle. Agron J 99:1252-1259.

Gosling, P., and M. Shepherd. 2005. Long-term changes in soil fertility in organic arable farming systems in England, with particular reference to phosphorus and potassium. Agric. Ecosyst. Environ. 105:425–432.

Graham, P.H. and C. P. Vance. 2003. Legumes: Importance and Constraints to Greater Use. Plant Physiol. 131: 872-877

Habash D.Z, S. Bernard, J. Shondelmaier, Y. Weyen and S.A. Quarrie. 2006. The genetics of nitrogen use on hexaploid wheat: N utilization, development and yield. Theoretical and Applied Genetics 114, 403–419.

Haq, M.U., A.P. Mallarino. 2005. Response of soybean grain oil and protein concentrations to foliar and soil fertilization. Agron. J. 97: 910-918.

Herencia, J. F., J. C. Ruiz-Porras, S. Melero, P. A. Garcia-Galavis, E. Morillo and C. Maqueda. 2007. Comparison between Organic and Mineral Fertilization for Soil Fertility Levels, Crop Macronutrient Concentrations, and Yield. Agron J 99:973-983.

Hirel, B., J. L. Gouis, B. Ney. A. Gallais. 2007. The challenge of improving nitrogen use efficiency in crop plants: towards a more central role for genetic variability and quantitative genetics within integrated approaches. Nutrition spl issue June 7. page 1-19.

Hochmuth, G.J. 2003. Progress in mineral nutrition and nutrient management for vegetable crops in the last 25 years. HortSci 38:999-1003.

Howard, D.D., C.O. Gwathmey, C.E. Sams. 1998. Foliar feeding of cotton: evaluating potassium sources, potassium solution buffering, and boron. Agron. J. 90: 740-46.

Hussain, S.A,. A blue print for maximizing yields of sugarcane. Sind Sugar Corp., Ltd, Karachi 4, 1982.

Kadir, S.A. 2004. Fruit quality at harvest of "Jonathan" apple treated with foliarly-applied calcium chloride. J Plant Nutr. 27: 1991-2006.

Kannan, S. 1990. Role of Foliar Fertilization on Plant Nutrition. In "Crops as Enhancers of Nutrient Use". Academic Press, Inc. Orlando, FL. pp. 313-348.

Kichey, T., Hirel B, Heumez E, Dubois F, Le Gouis J. 2007. Wheat genetic variability for post-anthesis nitrogen absorption and remobilisation revealed by ^{15}N labelling and correlations with agronomic traits and nitrogen physiological markers. Field Crops Research 102, 22–32.

Lambers, H., M.W. Shane, M.D. Cramer, S.J. Pearse and E.J. Veneklaas. 2006. Root structure and functioning for efficient acquisition of phosphorus: Matching morphological and physiological traits. Ann Bot 98:693-713.

Lewis, L. N., A vital Resource in Danger, Calif. Agriculture, Oct 1984 p 2.

Li, J., Z. Li, X. Liu, W. Zhou, J. Sun and Y. Tong. *et al.* 1995. Technique of wheat breeding for efficiency utilizing soil nutrient elements. *Science in China* (*Series B*) 38:1313-1320. (*see* Yan *et al.* 2006).

Lian, X., S. Wang, J. Zhang, Q. Feng, L. Zhang D. Fan. *et al.* 2006. Expression profiles of 10,422 genes at early stage of low nitrogen stress in rice assayed using a cDNA microarray. Plant Molecular Biol 60:617-631 (*see* Yan *et al.* 2006).

Logan, T.J. 1987. An assessment of Great Lakes tillage practices and their potential impact on water quality. In T.J. Logan, J.M. Davidson, J.L. Baker, and M.R. Overcash, eds. Effects of Conservation Tillage on Groundwater Quality. Lewis Pub. Chelsea, MI. Pp 271-276.

Lopez-Bellido, RJ., Shepherd CE, Barraclough, PB. 2004. Predicting post-anthesis N requirements of bread wheat with a Minolta SPAD meter. European J Agron 20, 313–320.

Mansfield, T.A., K.W.T. Goulding and L.J. Sheppard. 1998. Major biological issues resulting from anthropogenic disturbance of the nitrogen cycle. (The Third New Phytologist Symposium, Lancaster University 3-5 September 1997). New Phytol 139:1-2.

Mayer, P.M., S.K. Reynolds, Jr., M.D. McCutchen and T.J. Canfield. 2007. Meta-analysis of nitrogen removal in riparian buffers. J Environ Qual 36:1172-1180.

Maynard, D.N. and D.N. Lorenz. 1979. Controlled release fertilizers for horti culture crops. In "Horticultural Reviews" (J. Jules, Ed.), pp 79-140. AVI Publishing company Inc., Westport.

McGinnis, L. 2007. Show me the money: Why economics is essential for sustainable agriculture. Agricultural Research July, pp 8-10.

Mitchell, J.P., D.S. Munk, R. Prys, K.K. Klonsky, J.F. Wroble and R.L. de

Moura. 2006. Conservation tillage production systems compared in San Joaquin Valley cotton. Calif Agriculture 60, 3, 140-145.

Nakao, Seiji. 1998. Soil Conservation on Sloping Orchards in Japan, Food & Fertilizer Tech Center Ext. Bull 451.

Nelson, K.A., P. P. Motavalli, M. Nathan. 2005. Response of No-Till soybean [*Glycine max* (L.) Merr.] to timing of preplant and foliar potassium applications in a claypan soil. Agron. J. 97: 832-838.

Patni, N.K., L. Masse, and P.Y. Jui. 1998. Groundwater quality under conventional and no tillage: I

Nitrate, electrical conductivity, and pH. J. Environ. Qual. 27:869-877.

Peng S, Garcia FV, Laza RC, Sanico AL, Visperas RM, Cassman KG. 1996. Increased N-use efficiency using a chlorophyll meter on high yielding irrigated rice. Field Crops Research 47, 243–252.

Peuke, A.D., W.D. Jeschke, W. Hartung. 1998. Foliar application of nitrate or ammonium as sole nitrogen supply in *Ricinus communis*. II. The flow of cations, chloride and abscisic acid. New Phytol. 140: 625-636.

Ragothama, K.G. 1999. Phosphate acquisition. Annu Rev Plant Physiol Plant Mol Biol 50: 665-693.

Roe, E. 1998. Compost utilization for vegetable and fruit crops. HortScience 33:934–937.

Runge-Metzger, A. 1995. Closing the cycle: obstacles to efficient P management for improved global security. *In* H Tiessen, ed, Phosphorus in the Global Environment. John Wiley and Sons Ltd, Chichester, UK, pp 27-42

Schactman, D.P., R.J. Reid, S.M. Ayling.1998. Phosphorus uptake by plants: from soil to cell. Plant Physiol 116: 447-453.

Seta, A.K., R.L. Blevins, W.W. Frye, and B.J. Barfield. 1993. Reducing soil erosion and agricultural chemical losses with conservation tillage. J. Environ. Qual. 22:661-665.

Shoji, S., J. Delegado, A.R. Mosier and Y. Miura. 2001. Controlled release fertilizers and nitrification inhibitors to increase nitrogen use efficiency and to conserve air and water quality. Comm Soil Sci Plant Anal 32:1051-1070.

Shwartz, M. 2007. Ancient coral reef tells the history of Kenya's soil erosion. Stanford University Public Release date 10th April 2007.

Smith, E.G., R.D. Knutson, C.R. Taylor and J.B. Penson. 1990. Impact of chemical use reduction on crop yields and costs. Texas A&M Univ,. Dept Agric Econ., Agric and Food Policy Cent., College Station. (See Stewart *et al.*, 2005)

Steingrobe, B. 2001. Root renewal of sugar beet as a mechanism of P uptake efficiency. J Plant Nutr and Soil Sci 164:533-539.

Steingrobe, B., H. Schmid and N. Classen. 2001. Root production and root mortality of winter barley and its implication with regard to phosphate acquisition. Plant and Soil 237:239-248.

Stelly, M., Moving Up the Yield Curve: Advances and Obstacles, ASA Special Publ 39, 1980, Amer Soc Agron, Soil Sci Soc Am., Madison, Wisconsin.

Stevens, W.B., A.D. Blaylock, J.M. Krall, B.G. Hopkins and J.W. Ellsworth. 2007. sugarbeet yield and nitrogen use efficiency with preplant broadcast, banded, or point-injected nitrogen application. Agron J 99:1252-1259.

Stewart, W.M., D.W. Dibb, A.E. Johnston and T.J. Smyth. 2005. The contribution of commercial fertilizer nutrients to food production. Agron J 97:1-6.

Tal, Alon. 2006. Combating desertification: Israel's ecological triumph. Jewish Press, November 15.

Tanaka, D.L., J.M. Krupinsky, M.A. Leibig, S.D. Merrill, R.E. Ries, J.R. Hendrickson, H.A. Johnson and J.D. Hanson. 2002. Dynamic cropping systems: An adaptable approach to crop production in the Great Plains. Agron J 94:957-961.

Teasdale, J. 2007. No shortcuts in checking soil health. Agricultural Research July 2007, pp 4-5.

Thomas, G.W., R.L. Blevins, R.E. Phillips, and M.A. McMahon. 1973. Effect of a killed sod mulch on nitrate movement and corn yield. Agron. J. 65:736-739.

Towery, D. 2002. National Crop Residue Management Survey. Conservation Tillage Data. Conservation Technology Information Center, W. Lafayette, IN. www.ctic.purdue.edu/CTIC/CRM.html.

Vance, C.P. 2001. Symbiotic nitrogen fixation and phosphorus acquisition. Plant nutrition in a world of declining renewable resources. Plant Physiol 127:390-397.

Veenstra, J. J., W.R. Horwath, J.P. Mitchell and D. S. Munk. 2006. Conservation tillage and cover cropping influence soil properties in San Joaquin Valley cotton-tomato crop. 146 Calif Agriculture 90, 3, 146-153.

von Uexküll, H.R. and E. Mutert.1995. Global extent, development and economic impact of acid soils. Plant Soil 171: 1-15.

Wallace, A. 1984. The next agricultural revolution. Commun Soil Sci Plant Anal 1984, 15:191-197

Wallace, A.1990. The Interacting nature of limiting factors including micronutrients on crop production: Getting high yields from present cultivars, Proc Internt'l Congress of Plant Physiol. New Delhi, India, 1990, vol 2, pp 1164-1168.

Wallace, A., and G. Wallace. 2003. Closing the crop yield gap through better soil and better management: The Law of the Maximum. Wallace Laboratories, 10523 Holman Avenue, Los Angeles, CA 90024, USA and, 365 Coral Circle, El Segundo, CA 90245, USA.

Warman, P.R. 2005. Soil fertility, yield an nutrient contents of vegetable crops after 12 years of compost or fertilizer amendments. Biol. Agric. Hortic. 23:85–96.

Warman, P.R., and K.A. Havard. 1998. Yield, vitamin and mineral content of organically and conventionally grow potatoes and sweet corn. Agric. Ecosyst. Environ. 68:207–216.

Welch, R.M., 1999. Towards a Greener Revolution: Creating More Healthful Food Systems. Agric Research, May 1999, p.2.

Williams, C.M. 2002. Nutritional quality of organic food: Shades of grey or shades of green? Proc. Nutr. Soc. 61:19–24.

Wittwer, S.H., Food production: Technology and the resource base. Science 1975, 188:579-584.

Woolfolk, C.W., W.R. Raun, G.V. Johnson, W.E. Thomason, R.W. Mullen, K.J. Wynn, and K.W. Freeman. 2002. Influence of late-season foliar nitrogen in winter wheat. Agron. J. 94: 429-434.

Wu, Chia-Chun. 1998. Food & Fertilizer Technology Center Extension Bull 449.

Wuest, S.B., K.G. Cassman. 1992. Fertilizer-nitrogen use efficiency of irrigated wheat. I. Uptake efficiency of preplant *vs.* late-season application. Agron. J. 84: 682-688.

Yan, X., P. Wu, H. Ling, G. Xu, F. Xu and Q. Zhang. 2006. Plant nutriomics in China: An overview. Ann Bot 98:473-482.

CHAPTER-III

Hard Facts And Realities of Food Production

HARD FACTS AND REALITIES OF FOOD PRODUCTION

The United Nations released a report on the World Food Day in 1999 that nearly 80% of all malnourished children in the developing world in the early 1990s lived in countries that claimed food surpluses, which means that people often go hungry even though food is readily available. The economists put forth the arguments that poverty rather than food shortages is the real cause of hunger. In the hungry countries it is the pervasive poverty which limits people's access to food. Poverty also prevents access to non-food services like health care, education and a clean environment. In addition, medical conditions like diarrhea, which generally results from the use of contaminated water supply, prevents a child from absorbing available nutrients from even the scanty food available to them.

World Food Day is celebrated every year on the 16th October. Since 1981 it has adopted a different theme each year in order tohighlight the areas for action. The theme for 2006 was "Invest in agriculture for food security". The years 1981 and 1982 were for "Food comes First". It is over 25 years and still this theme is unsatisfied and has to be considered of prime importance. So also the themes for several years, viz., 1996 was "Fighting Hunger and Malnutrition", 2001 for "Fight Hunger to Reduce Poverty", 1994 for "Water for Life", 2002 for "Water: Source of Food Security" and this 2007 is "Right to Food". While it is interesting to know the topics of dedication for each year, it is a mute question to ask how much was accomplished in the year that was and those years following that year. Still the food production, fighting against hunger, and water requirement for life and agriculture are far from reaching any level of being proud of. While it does not mean that there should be 'no theme', it is important to analyse the past and made amends for the future. The theme for 2000 is for "A millennium Free from Hunger". Though it is seven years from that celebration, still the malnutrition and hunger continue to grow to a frightening level.

The USDA yearbook of Agriculture 1969 was devoted for the subject *Food for Us All*. Food for us all started with 3 million farmers who produced the animal and other foods consumed at the rate of 6000 pounds a year by a family of four, and about half of the total land area in the 50 states was used for farming. The farm operations were highly mechanized and the food for the Americans constituted only a small proportion of their take-home pay, the lowest of any country. Furthermore, the farming was so efficient that one farm worker supplied products for about 43 people. More than 23 million people were in agri-business which was 30% of the work force. School children were supplied milk and in some states as in Georgia, they were given lunch too, with a view to improve the nutrition of children and the poor. In the foreword to the yearbook Secretary of Agriculture Clifford M. Hardin stated "Food for us. ...Means enough food—and the right food—for every American. In a nation with the greatest food production capacity ever achieved—an ability in fact to produce considerably more than domestic and foreign markets can absorb—one might assume that the goal of food for us all would long since have been realized. Yet, it has not. There is still hunger in the midst of national plenty. Poverty-caused malnutrition is far more widespread than we previously had thought. Under-nutrition, caused by ignorance or neglect, is a common fact of American life. There are still great gaps to be filled in the science of nutrition. The campaign to educate people about what makes a good diet and the relationship between diet and vigor, health, and longevity is only in its beginnings.........He (farmer) wants to produce enough to meet the nutritional needs of every man, woman, and child in this country....He does, and more....As consumers, we have a stake in productive and prosperous agriculture......Farm people are not sharing adequately in the Nation's abundance. They have less than three-fourths as much income per person as non-farm people". Though there is a general impression that Americans have no problem as far as food production or consumption is concerned, Hardin tells a different story. Though there is no shortage of food in America, there is malnutrition and under-nutrition which continue because there is lack of income amongst these people who are too poor to buy an adequate diet from the abundance of food in the commercial market. In addition, many of these poor people are not knowledgeable with regard to simple basic dietary rules, and the importance of good nutrition.

Revelle in 1974 made a comparison of the total food intake in terms of calories by Indians and Americans. Rice taken by person in India are nearly 30 times that as taken by an American who takes meat 80 times more than the Indians. It is interesting to look at the developments in the nutrition of Americans almost 20 years later, a situation described by Enns, Tippett, Basiotis and Goldman. "Americans are certainly eating more grains. The average intake of grain products increased from 213 gms in 1977-78 to 254 gms in 1989-90—a 19% increase.....Americans ate about the same amount of fruit in 1989-90 as in 1977-78, but some people are eating little or no fruit. More than a fourth of the population ate no fruit and drank no fruit juice during the 3 consecutive days of record-keeping....A larger proportion of low-income people (33%) ate no fruit than did high-income people (23%)....In 198-90, Americans drank about the same amount of total milk and milk products as they did in 1977-78. As a share of total milk and milk products, however, low fat and skim milk went up while whole milk went down".

Many scientists have discussed agricultural Research at the dawn of the Millennium and the research in the Agricultural Research Stations in the US in the past, and the present as also the projections for the future. (*see* Agricultural Research

December issue of 1999 for these authors). Walter H. Wischmeier who worked with USDA during 1940-1976 developed techniques for evaluation of soil erosion and according to him local participation in the management of research are essential in order that the technology developed is fully utilized. R. James Cook (at USDA 1965-1998) introduced biological control for plant pathogens and the growing of wheat with reduced tillage. He stated "High-yield agriculture has long been associated with environmental devastation. But through genetics and management, we've moved closer to obtaining these yields with benign environmental consequences". "It is a temptation to adopt new methods just because they are high-tech, even though they may not have been shown to be better and more economical than traditional methods." (Crittenden, at USDA 1961–1989).

The Yearbook of Agriculture of the year 1962 is devoted to the description of events after a hundred years in the US, as the title goes. In the Foreword, O.L. Freeman, Secretary of Agriculture described about the achievements in the previous hundred years which cannot be measured in tons or dollars or even in terms of stomachs filled and bodies clothed. It was also a period of challenges met and responsibility laid upon the people. American farmers produced many times more than in the hundred years before. The number of americans who were ill fed and ill clothed were reduced remarkably. The technological Revolution now is far beyond the Industrial Revolution in scope and possibilities. The americans have enough food in storage to see them through any emergency. But in the coming century the responsibility is much higher than before. And people must march confidently beyond if they are to survive and the free world is to survive.

Alfred Stefferud, the editor of the yearbook summarized the achievements during the century ending 1962. The farm population continued to decline in size, due to the great recent advances in technology, but productivity has not reached their peak. Farm operation has now become a matter of sophisticated management. American farm people have a strong influence on the whole population. And so, after a hundred years, people have reached a moment in history that gives them the opportunity for rededication. The americans have traditions and history and from them—facts and precepts for guidance in the years to come. Science and technology that expand beyond our ability to see their end; an awareness of the universe and our part in it viz., the land, rivers, lakes, mountains, forests, deserts, plains, valleys, space and people. We have a legacy. The legacy of American farmers sets new goals and responsibilities to insure a fair return to farmers, provide basic needs and care for all the resources in the planet.

2. CONCLUDING REMARKS

From the many of the reports described a few important points are evident. There is no second opinion about the fact that population is increasing greatly and food production will be behind the race. Population has to be controlled. In this millenium, the rate of growth could at least be reduced by adopting family planning measures. It is also possible to augment food production. Revelle (1974) has brought out one important point, "When Thomas Malthus announced in 1798 his famous Principle of Population....he thought that the resources available for agriculture were primarily land, water and human and animal labor". But progress in agriculture 150 years after his Malthus's statement, has been unprecedented that Malthus would not have foreseen. Research in plant breeding, improvement of soil fertility, improvement of the photosynthetic efficiency, evolving of crop cultivars which are responsive to high fertilizers, vast increase in the irrigation systems all ushered in the Green Revolution. Feeding the billions thus became a reality in this century. What next and more are the questions to be tackled in the coming decades. Genetically modifying the present high-yielding crop cultivars is a very fruitful area. However, the inputs necessary for the GM plants to manifest the effects are to be provided. As an example, if a new GE potato which can produce more proteins are introduced, the ingredients like nitrogen, phosphorus and other elements are to be made available through the soil. Thus soil fertility is still important. Furthermore, such GM potato can make more proteins, but may not have the capacity to effectively absorb the nutrients from soil or air, or even may not be efficient in photosynthesis. The subject of genetic engineering has been reviewed and its role is compared to that of crop domestication and conventional plant breeding by Gepts (2002). He has made a few conclusions. He does not believe that changes brought in through GE would make the crops 'weedier'. However the GE would bring in gain-of-function mutations in contrast to the 'loss-of-function' from natural mutations in domestication.

Introduction through innovations as in GE does not automatically improve the well being of people. While GE is not a substitute for conventional plant breeding, it can bring in additional genetic materials. He advocates case-by-case assessments of the benefits and losses before introduction of GE techniques for crop improvement. His recommendation is that there should be impartial discussion with the scientists and the public on the issues involved in genetic engineering of crop plants, for the benefit of the people.

The editorial in the California Agriculture, July-Sept 2006, discusses the present changes in global climate and other ecological and socioeconomic crises, looming large in the horizon. Global climate change will lead to adaptations to local conditions and crop varieties. Farmers in the U.S. also are facing threats from international and domestic markets.Just as most oil production takes place abroad, ammonia-based fertilizers and potassium and phosphorus supplies will also be affected. All of these issues create conditions for quick biotechnological solutions to help farmers adapt and conserve precious resources. The world's population will likely increase by about 50% in the next 50 years, along with an increased standard of living. These trends will increase the demand for food, fiber and energy. To meet this demand, U.S. agriculture faces a situation similar to the period of transition when plant breeding and synthetic fertilizers led to new heights in farm production. To meet this new crisis, there will likely be a transition to genetically modified crops to suit with a variety of input and output traits.

Just like traditional crops, genetically engineered crops could occasionally create the same problems. Ellstrand (2006)describes about the problems with transgenic plantgs thus: "Currently in California, the movement of transgenes from most commercialized transgenic crops into wild plant populations is unlikely — the exception being canola. However, other transgenic plants have been field-tested in California, and if these become commercialized, in certain cases, transgenes are likely to move into the wild or into other crops of the same species. Such gene flow could result in various problems."

Soil and water sources are not inexhaustible. But considerable efforts should be taken to increase the area of farm land and the irrigation sources, without affecting their quality. There are many examples where this has been achieved and have further scope for expansion (Avery, 1985). Some nations such as Sudan, Zimbabwe and Thailand have large areas of uncropped land which are however having limitations of inapproachability and transport. Turkey is building large dams to irrigation millions of hectares in the upper Euphrates valley. Brazil is opening up 50 million hectares of acid soils using lime and phosphate fertilizers. In India, Australia and the Sudan, about 300 million hectares of black vertisol soils are being reclaimed and brought under cultivation. Pasture lands are shifted to intensive cropping with grain and oilseeds in Argentina. In Peru, a new upland variety of rice tolerant to Al is introduced in land having excess Al. Soil erosion which reduces the good soil, is prevalent in many countries. Conservation tillage and minimum tillage techniques are introduced in many countries. Africa has the worst of soil erosion problem. This is largely because most of the food production is carried out in millions of tiny farm lands. Overgrazing also increases erosion. Better conservation measures and improved production technologies would surmount many of the African food problems. A new sorghum hybrid is introduced in Sudan and a new sorghum suitable for drier regions in Sahel lead to better trends towards self-sufficiency. A leguminous tree native of Central America, *Leucaena leucocephala* is very useful for Africa, and it can supply timber as well as prevent soil erosion. However, according to Avery, the most serious constraint on African agriculture is the political systems which have national policies not in favor of concentrated and long-term efforts toward self-sufficiency in food.

The goal of food production is manifolds. Self-sufficiency in supplying food for the people is one. Providing good quality and nutritious food is equally important. Another front is controlling common diseases arising out of poor quality food, contaminated water and lack of fresh air. Above all, minimal requirement of these should be guaranteed to the mentally and physically handicapped men, women, and especially children, all of them anxiously looking upon the otherwise well-to-do society and the nation as a whole.

Walker and Steffen (1997) have discussed the effects of global change which is seen in the land use, atmospheric composition and biological diversity and climate as revealed from the studies under the project IGBP (International Geosphere-

Biosphere Programme). The elevated CO_2 in the atmosphere will have a positive effect with an increase of aboveground biomass upto 80% and the biosphere will turn a source for carbon and not a sink, in the next century. The authors forecast that the species present in the terrestrial biosphere would get impoverished. It is also said that an elevated CO_2 would likely increase crop production of the temperate regions, while the same will decline in the tropics at least in some regions. Since crop production has to be increased to at least 2% from the present one, conversion of pastures and rangelands would necessarily result. The efforts to increase the crop production will create stress on the management of production systems. The frequent droughts in a few grain-producing areas also would lead to considerable reduction in yield. In other words, producing more food and fiber will be affected by climate changes. At the same time, other resources needed for food production, like water would be in shortage, especially in view of the urban demands from people and industry which is taking place in the developing countries. The authors summarized the effects of global change thus: Crop production will be affected differently in different countries, and will be decreased in countries at the low latitudes where the food demand will be on the increase. Wheat yield will be on the increase upto 10% while rice yield would decrease by 5% with a 1° rise in temperature above $32^{\circ}C$.

The Year book of Agriculture for 1991 is fully dedicated to the Environment. The Secretary of Agriculture Edward Madigan states "From the Dust Bowl days of the 1930's to the President's Water

Quality Initiative of today, USDA has been charged with taking positive steps to help protect America's lands and improve water quality". Environmental quality and economic growth must go together. The current level of global food production was attained with the use of chemical inputs, mechanical power, and irrigation. New technologies, developed through both government and private research, have greatly increased the productivity of land. And new technologies will further help attain agricultural production goals. Discussing about sustainable agriculture, O'Connell in the yearbook 1991 states that the USDA's agricultural research service spends millions of dollars annually for research in sustainable agriculture. About $83 million is devoted annually for research on biological pest control and integrated pest management, evolving of crop varieties resistant to many important pests and disease, and those resistant to drought, adverse soil conditions, and make farmers less dependant on agricultural chemicals like pesticides and increasing crop productivity, and so many other important disciplines.

Sinclair *et al.* (2004) have discussed the role of molecular transformation of plant traits in order to overcome the yield plateau reached in crops. While it was expected that genetic engineering involving gene transfer could raise the yield potential, it has not been successful in all cases excepting in the introduction of traits for pest and disease control. For developing countries, breaking the yield barrier is the top priority. It is therefore necessary to look at the possibility of gene transfer for physiological traits which could increase yield. The most important one is to increase photosynthetic capacity. Attempts were made to introduce C_4 pathway for C_3 species and transgenic rice lines having higher levels of PEPcaraboxylase activity; but the photosynthetic rates of such plants were not higher than the normal ones. Introduction of C_4 pathway without the associated morphological changes led to the failure. Another factor, nitrogen has been crucial for the yield increase obtained earlier for several years. Nitrogen is essential for the grain formation and increased protein synthesis. Rubisco is a primary storage site for nitrogen during plant vegetative development, and the grain development begins with the Rubisco breaking down and releasing the nitrogen for transport to the grain. Nitrogen is required to primarily synthesize photosynthetic enzymes and the other nitrogenous compounds. If nitrogen is readily available in the soil this increase can be met with readily, and then sufficient nitrogen will have been stored in Rubisco to allow a higher yield. To achieve this, the genetically transformed plant will have to have an enhanced nitrogen storage mechanism. But if this mechanism is not introduced, the calculated yield increase can become negative. Therefore, the larger vegetative plants that form as a result of stimulated photosynthesis mean that more nitrogen needs to be incorporated into structural components of the vegetative tissue leaving no nitrogen available for transfer to the grain. Thus additional nitrogen is required for transport to the grain. The goal of molecular genetic researchers to increase 'nitrogen use efficiency' means nothing in terms of crop yield. Next approach would be to increase the growth rate of individual seeds to increase overall grain production. A large amount of past research has focused on this and it was found that attempts to improve the seed growth did not increase harvest index. Another line of research relating to the transformation of ADP-glucose pyrophosphorylase in seed, so that there is less sensitivity to phosphorus inhibition, has been pursued in order to sustain and increase seed growth rate. Although increased seed growth of the genetically transformed wheat and rice was reported, it increase only the growth of the plants and did not lead to an increase in the plant harvest index. The authors concluded that for a successful program to develop improved cultivars, at least ten years of research is needed and a sustained commitment of scientists to apply molecular understanding of the objective of increasing crop yield. The funding agents should favor long-term commitments for supporting research.

Finally there is an important question. In the last 50 years, we have observed and brought forth several changes due to necessities and inventions. Farmers had been using organic manuring for a long time, but inorganic fertilizers were produced and extensively used to boost crop yields. In the last two decades, these were considered excessive and harmful. So we look back to organic fertilizers. We went far ahead in controlling pests and diseases, using newly discovered organic chemicals which now are contaminating the soil and biological systems. As a result now we look for harmless biodegradable pesticides or alternative measures. The question is, will we ever look back to conventional breeding, from the present trend of introducing GE plants? A symposium was held at

Michigan State University in 2005 and the future of plant breeding education was discussed (Gepts and Hancock, 2006). The support for plant breeding program in the universities in the USA and other developed countries as well as the International Agricultural Research Centers seems to be on the decline (Knight, 2003). This is of great concern to the plant breeding industry in general. The major themes discussed in the symposium were: Define plant breeding, describe plant breeding education and employment, address the critical needs of breeding as a specialty and create its awareness. Plant breeding is defined as a multidisciplinary science, which involves the application of genetic principles, agronomy, botany, molecular genetics, physiology and biochemistry amongst other fields. In the last few decades, it is specialized in using technologies like transfer of genes having certain quantitative traits leading to increased plant productivity. While the traditional plant breeding relates to the development of new cultivars, its definition is inadequate and the present view is that it should include crop improvement through breeding research.

Finally it is worth quoting what L.H. Bailey stated nearly a century ago:

3. "EVERY FARMER SHOULD BE AWAKENED": LIBERTY HYDE BAILEY'S VISION OF AGRICULTURAL EXTENSION WORK"

(Scott Peters; 2006. American History, 80:2:190-219).

Historians have portrayed the formative period of agricultural extension work in the United States as a search for the best method of convincing farmers to change their farming practices in order to improve agricultural efficiency, productivity, and profitability. However, one of the key leaders in extension's formative period, Cornell University's Liberty Hyde Bailey, articulated a different vision of extension's central purpose and promise. Drawing on his writings during the years in which he led the development of Cornell's extension program (1894-1902), this article argues that Bailey's vision of agricultural extension work was centered on the provision of education aimed at awakening farmers to a new point of view on life. The new point of view combined sympathy with nature, a love of country life, and a scientific attitude, expressed by a habit of careful observation and experimentation. The main purpose of awakening farmers to this point of view was not to develop a more efficient, productive, and profitable agriculture, but to advance the larger cultural ideals of a "self-sustaining" agriculture and personal happiness. The account of Bailey's vision provided in this article suggests the need to reconsider the story of the origins and early development of American agricultural extension work.

It is very often stated that the developing countries should plan for a second Green Revolution. It is important to realize revolutions do not take place by 'ups' and 'starts' nor there is a 'quantum leap' with a distinctive 'breaks' between the revolutions. The history of agriculture described in the beginning clearly shows the so-called 'revolutions' have been taking place throughout the various stages from the time humans evolved themselves from 'foragers' and 'hunters' to 'farmers';

furthermore, domestication of plants by humans itself is the first revolution, that took place thousands of years ago. The name 'Green Revolution' was first applied in March 1968 by the then administrator of Agency for International Development, William S. Gaud in his talk to the Society in Washington DC. Gaud said "These and other developments in the field of agriculture contain the making of a new revolution. It is not a violent Red Revolution like that of the Soviets, nor is it a White Revolution like that of the Shah of Iran. I call it the Green Revolution". The phenomenal success in crop yields after their introduction in the developing countries has been responsible for saving people from starvation. Furthermore, this success was followed by improvement of agriculture as a whole, with the enormous flow of money for the same, from the respective governments and international agencies. The fact should not be forgotten that the end of the second World War brought in lot of advancements like fertilizer manufacture, improved irrigation facilities and modern methods of pest and disease control not only for plants but also for humans. Before the war, however, plant breeding and improved crop cultivation were nonetheless pursued, at a slower pace lacking the momentum derived after the end of war. Starvation of millions of people provided the impetus for developing agriculture.

REFERENCES

Agricultural Research 1999. December issue USDA: Top Scientists Point to Research Past, Present and Future. R. James Cook (at USDA 1965–1998); Lyman B. Crittenden (at USDA 1961–1989) Walter H. Wischmeier (at USDA 1940–1976).

Avery, D. 1985. U.S. Farm Dilemma: The global bad news is wrong. Science 230:408-412.

Ellstrand, N.C. 2006. When crop transgenes wander in California, should we worry?. California Agriculture 60:3:116-117.

Gepts, P. 2002. A comparison between crop domestication, classical plant breeding and genetic engineering. Crop Sci. 42:1780–1790.

Gepts, P. and J. Hancock. 2006. The future of plant breeding—(Opinion and policy). Crop Sci 46:1630-1634.

Knight, J. 2003. Crop improvement: A dying breed. Nature 421:568–570

Revelle, Roger. 1974. Food and Population, Scientific American 231:161-170

Sinclair, T.R., L.C. Purcell and C.H. Sneller. 2004. Crop transformation and the challenge to increase yield potential. Trends in Plant Science 9:71-75.

Walker, B., and W. Steffen. 1997. An overview of the implications of global change for natural and managed terrestrial ecosystems. Conservation Ecology [online]1(2): 2. Available from the Internet. URL: http://www.consecol.org/vol1/iss2/art2

INDEX

--*